ARITHMETIC MODULE SERIES

Module 2: FRACTIONS

THOMAS J. McHALE
PAUL T. WITZKE
Milwaukee Area Technical College, Milwaukee, Wisconsin

ADDISON-WESLEY PUBLISHING COMPANY
Reading, Massachusetts
Menlo Park, California · London · Amsterdam · Don Mills, Ontario · Sydney

ARITHMETIC MODULE SERIES

Programmed Modules
Module 1: Whole Numbers
Module 2: Fractions
Module 3: Decimal Numbers
Module 4: Percent, Ratio, Proportion
Module 5: Rounding, Estimation, Squares and Square Roots, Formula Evaluation, Exponents and Scientific Notation

Nonprogrammed Text
A single text with the content of all five modules in shorter, nonprogrammed form.

Reproduced by Addison-Wesley from camera-ready copy prepared by the authors.

Second printing, May 1976

Copyright © 1975 by Addison-Wesley Publishing Company, Inc. Philippines copyright 1975 by Addison-Wesley Publishing Company, Inc.

All rights reserved. No part of this publication may be reproduced, stored in a retrieval system, or transmitted, in any form or by any means, electronic, mechanical, photocopying, recording, or otherwise without the prior written permission of the publisher. Printed in the United States of America. Published simultaneously in Canada.

ISBN 0-201-04752-7
RST-MU-943210

Preface

The Arithmetic Module Series is designed to teach the basic skills of arithmetic on an individualized basis. Developed by the co-authors of the MATC Mathematics Series, the Arithmetic Module Series presently includes the modules listed on the opposite page. To efficiently meet the needs of two distinct types of students - those who need a significant amount of original learning and those who need more of a review - the modules are available in both a programmed form and a shorter non-programmed form. The programmed modules are available only separately; the non-programmed modules are available only in a single text. Though both forms were designed for individualized instruction, the non-programmed form can easily be used in conjunction with the traditional lecture method.

The modules are accompanied by the booklet Tests For Arithmetic Module Series. That booklet contains the following types of tests: (1) diagnostic pre-tests for each module; (2) assignment diagnostic tests; (3) unit tests; (4) multi-unit tests; (5) comprehensive tests for each module; and (6) a general comprehensive test for all modules. Three equivalent forms are provided for all multi-unit and comprehensive tests and for a few special unit tests.

> Note: The test book is provided only to teachers, not to individual students. Copies of the tests for student use must be made by the Xerox process or some similar process.

The Series was developed and field tested over a two-year period in the junior college Math Workshop at MATC. In the Math Workshop, the full set of modules is used in a one-semester arithmetic course for students in both transfer and career programs. Individual modules are used as remedial materials for students referred from other math courses. In addition to the field testing at MATC, the modules were also field tested in some experimental classes in high schools in the Milwaukee area. Therefore, the Series can be used at both the college or high school level either as a basic course or as supplementary materials in other math courses.

In the spirit of the CUPM report "A Course in Basic Mathematics for Colleges," the Series is both terminal and preparatory. It is terminal because the modules emphasize the basic arithmetic skills needed for daily living and many rudimentary careers. It is preparatory because the modules include sufficient "understanding" to prepare students for elementary algebra and more advanced career math courses.

The Series fits the growing trend toward individualization in remedial courses in three ways. First, the availability of the modules in both a programmed and a non-programmed form provides the opportunity to prescribe either an "original learning" or a "review" type of instruction for each student. Second, the diagnostic pre-tests can be used to prescribe an individualized course content for each student by identifying the specific units within each module in which the student is deficient. Third, the instruction itself can be individualized because each student can proceed at his or her own pace. This type of individualizing is not only flexible and efficient, but effective. For example, in the Math Workshop at MATC, the average percent score on tests ranges from the mid 80's to the low 90's.

The authors wish to thank Dr. George A. Parkinson, Director Emeritus of MATC, and Dr. William L. Ramsey, Director of MATC, who made the development of this Series possible; Keith J. Roberts and Allan A. Christenson, mathematics instructors at MATC, who were involved in the field testing and offered many constructive suggestions; Ms. Arleen D'Amore and Ms. Marylou Hansen, who typed the camera-ready copy; Ms. Julie S. McHale, who proofread the camera-ready copy.

Prerequisites: A knowledge of whole-number operations with smaller numbers is assumed. The content of Module 1: Whole Numbers is more than adequate for this purpose.

Sequencing: The units should be covered in their numerical order. Unit 6 is an optional unit.

Features: Though adequate attention is given to "understanding", the major emphasis and goal of the module is to develop skills in the four basic operations with fractions and mixed numbers.

In order to emphasize the meaning of fractions and the principles needed in operations with them, the terms in the fractions are generally restricted to numbers smaller than 100.

Instead of including topics like divisibility, multiples, factoring, prime and composite numbers, and prime factors in a preliminary unit, they are taught as needed.

To find a lowest common denominator, the direct method of testing multiples of the larger denominator to find the lowest common multiple is emphasized. The treatment of the prime factoring method is delayed until a final unit, and that unit is optional.

The cancelling process is taught as a way of simplifying multiplications.

Special sections are devoted to contrasting the four operations for both fractions and mixed numbers.

All possible operations involving a whole number and either a fraction or a mixed number are discussed.

The fact that a fraction stands for a division is emphasized.

How To Use Module 2: Fractions

The procedure for using Module 2: Fractions is outlined below. Except for the assignment self-tests, all tests are provided in Tests For Arithmetic Module Series (available from the Addison-Wesley Publishing Company, Reading, Massachusetts 01867).

1. The results on the diagnostic pre-test for fractions can be used to prescribe an individualized program for each student. The "Individualized Program - Fractions" sheet available in Tests For Arithmetic Module Series can be used to identify for each student the assignments to be completed and the completion date for each assignment.

2. After a student has completed each required assignment and the assignment self-test (in the module), the assignment diagnostic test can be administered. The assignments for Module 2: Fractions are listed at the bottom of this page and on the "Individualized Program - Fractions" sheet. Since the assignment diagnostic tests are designed to take only 15 or 20 minutes, ample time should be available for immediate correction and tutoring. These tests are simply a teaching tool and need not be graded.

3. After the appropriate assignments are completed, either a unit test or a multi-unit test can be administered. Ordinarily, these tests should be graded.

    Note: (a) In the Math Workshop at MATC, only multi-unit tests are administered on a regular basis in order to make the instruction more efficient.

    (b) Unit tests are administered only in special cases at the discretion of the instructor, with the exception that students who omit Unit 6 are given the Unit 5 test instead of the Multi-Unit Test: Units 5 and 6.

4. After all prescribed assignments are completed in the manner above, the comprehensive test for Module 2: Fractions can be administered. Since the comprehensive test is equivalent to the diagnostic pre-test for Module 2: Fractions, the difference score can be used as a measure of each student's improvement.

ASSIGNMENTS FOR MODULE 2: FRACTIONS

Unit 1: #1 (pp. 1-8)
        #2 (pp. 9-17)

Unit 2: #3 (pp. 18-28)
        #4 (pp. 28-38)

Unit 3: #5 (pp. 39-47)
        #6 (pp. 48-58)
        #7 (pp. 58-66)

Unit 4: #8 (pp. 67-75)
        #9 (pp. 76-86)
        #10 (pp. 86-94)

Unit 5: #11 (pp. 95-102)
        #12 (pp. 103-110)
        #13 (pp. 111-119)

*Unit 6: #14 (pp. 120-129)
         #15 (pp. 129-136)

---

*Optional unit

C O N T E N T S

UNIT 1   THE MEANING AND EQUIVALENT FORMS OF PROPER FRACTIONS   (pages 1-17)

    1-1   Using Fractions To Compare A Part With A Whole   1
    1-2   The Meaning Of The Words "Numerator" and "Denominator"   2
    1-3   Word Names For Fractions   3
    1-4   Pairs Of Equivalent Fractions   4
    1-5   Families Of Equivalent Fractions   5
    1-6   Raising Fractions To Higher Terms   7

    Self-Test For Assignment 1   8

    1-7   The Concept Of Divisibility   9
    1-8   Special Tests For Divisibility By 2, 3, And 5   10
    1-9   Reducing Fractions To Lower Terms   12
    1-10 Reducing Fractions To Lowest Terms   13
    1-11 Determining Whether Two Fractions Are Equal Or Not   16

    Self-Test For Assignment 2   17

UNIT 2   THE MEANING AND EQUIVALENT FORMS OF IMPROPER FRACTIONS AND MIXED NUMBERS   (pages 18-38)

    2-1   Fractions That Equal The Number "1"   18
    2-2   Fractions In Which The Numerator Is Larger Than The Denominator   19
    2-3   The Meaning Of Proper And Improper Fractions   20
    2-4   Raising Improper Fractions To Higher Terms   20
    2-5   Reducing Improper Fractions To Lowest Terms   21
    2-6   Fractions That Equal Whole Numbers Larger Than "1"   22
    2-7   Fractions Whose Denominators Are "1"   24
    2-8   Converting Whole Numbers To Fractions   24
    2-9   Fractions Whose Numerators Are "0"   26
    2-10 Fractions On The Number Line   27

    Self-Test For Assignment 3   28

    2-11 Addition Of "Like" Fractions   28
    2-12 The Meaning Of Mixed Numbers   30
    2-13 Converting Mixed Numbers To Improper Fractions   31
    2-14 Converting Improper Fractions To Mixed Numbers   33
    2-15 Mixed Numbers On The Number Line   35
    2-16 Relating Fractions And Division   37
    2-17 Writing Any Quotient With A Remainder As A Mixed Number   38

    Self-Test For Assignment 4   38

UNIT 3 ADDITION AND SUBTRACTION OF FRACTIONS   (pages 39-66)

- 3-1 Adding Fractions With "Like" Denominators   39
- 3-2 Subtracting Fractions With "Like" Denominators   40
- 3-3 The Meaning Of Multiples   42
- 3-4 Adding Fractions When The Larger Denominator Is A Multiple Of The Smaller   43
- 3-5 Subtracting Fractions When The Larger Denominator Is A Multiple Of The Smaller   46

  Self-Test For Assignment 5   47

- 3-6 Using Families Of Equivalent Fractions For Additions In Which The Larger Denominator Is Not A Multiple Of The Smaller   48
- 3-7 The "Lowest Common Denominator" Concept   50
- 3-8 A Strategy For Identifying Lowest Common Denominators   52
- 3-9 Adding Fractions When The Larger Denominator Is Not A Multiple Of The Smaller   55
- 3-10 Subtracting Fractions When The Larger Denominator Is Not A Multiple Of The Smaller   56

  Self-Test For Assignment 6   58

- 3-11 Adding Three Fractions   58
- 3-12 Using The Product Of The Denominators As The Common Denominator   61
- 3-13 Comparing The Size Of Fractions   63
- 3-14 Applied Problems   65

  Self-Test For Assignment 7   66

UNIT 4 MULTIPLICATION AND DIVISION OF FRACTIONS   (pages 67-94)

- 4-1 The Procedure For Multiplying Two Fractions   67
- 4-2 Justifying The Procedure For Multiplying Two Fractions   69
- 4-3 Multiplications Involving A Fraction And A Whole Number   69
- 4-4 Multiplications Involving A Fraction And Either "1" Or "0"   71
- 4-5 Using Parentheses As The Multiplication Symbol   72
- 4-6 The "Cancelling" Process For Multiplications Involving Fractions   73
- 4-7 Applied Problems   75

  Self-Test For Assignment 8   75

- 4-8 Pairs Of Reciprocals   76
- 4-9 The Procedure For Dividing Fractions   77
- 4-10 Divisions Involving A Whole Number And A Fraction   79
- 4-11 Justifying The Procedure For Divisions Involving Fractions   80
- 4-12 Division Of Fractions In Complex-Fraction Form   81
- 4-13 Divisions Involving A Fraction And Either "1" Or "0"   84
- 4-14 Applied Problems   85

  Self-Test For Assignment 9   86

- 4-15 Multiplying Three Fractions   86
- 4-16 The "Cancelling" Process For Multiplications Involving Three Fractions   87
- 4-17 Three-Factor Multiplications Involving Fractions And Whole Numbers   89
- 4-18 Contrasting The Four Operations With Fractions   91
- 4-19 Mixed Applied Problems   92

  Self-Test For Assignment 10   94

UNIT 5   OPERATIONS WITH MIXED NUMBERS   (pages 95-119)

- 5-1   Adding Mixed Numbers With "Like" Denominators   95
- 5-2   Adding Mixed Numbers With "Unlike" Denominators   96
- 5-3   Adding A Mixed Number To A Whole Number Or A Fraction   97
- 5-4   Mixed-Number Sums That Require A Regrouping   98
- 5-5   Adding Three Mixed Numbers   100
- 5-6   Applied Problems   102

     Self-Test For Assignment 11   102

- 5-7   Subtracting Mixed Numbers   103
- 5-8   Subtracting Mixed Numbers When "Borrowing" Is Needed   104
- 5-9   Subtracting Whole Numbers And Fractions From Mixed Numbers   106
- 5-10   Subtracting Mixed Numbers And Fractions From Whole Numbers   108
- 5-11   Comparing The Size Of Mixed Numbers   109
- 5-12   Applied Problems   109

     Self-Test For Assignment 12   110

- 5-13   Multiplications Involving Mixed Numbers   111
- 5-14   Three-Factor Multiplications Involving Mixed Numbers   112
- 5-15   Divisions Involving Mixed Numbers   113
- 5-16   Multiplications And Divisions Involving A Mixed Number And Either "1" Or "0"   115
- 5-17   Contrasting The Four Operations With Mixed Numbers   116
- 5-18   Estimating Answers To Mixed-Number Problems   117
- 5-19   Mixed Applied Problems   118

     Self-Test For Assignment 13   119

UNIT 6   PRIME FACTORING AND LOWEST COMMON DENOMINATORS   (pages 120-136)

- 6-1   The Need For Another Method For Identifying Lowest Common Denominators   120
- 6-2   Factors And The Factoring Process   121
- 6-3   Prime And Composite Numbers   123
- 6-4   The Meaning Of Prime Factoring   124
- 6-5   Prime Factoring By The Smallest-Prime Method   126
- 6-6   Prime Factoring By The Factor-Tree Method   127

     Self-Test For Assignment 14   129

- 6-7   Finding Lowest Common Denominators By The "Prime-Factoring" Method   129
- 6-8   A Shorter Form Of The Prime-Factoring Method   132
- 6-9   The Prime-Factoring Method And Additions Of Three Fractions   135

     Self-Test For Assignment 15   136

Unit 1 THE MEANING AND EQUIVALENT FORMS OF PROPER FRACTIONS

Proper fractions are fractions in which the numerator is smaller than the denominator. We will discuss the meaning and equivalent forms of proper fractions in this unit. The procedure for raising proper fractions to higher terms and the procedure for reducing proper fractions to lowest terms will be discussed. A special section is devoted to the word names for proper fractions.

1-1 USING FRACTIONS TO COMPARE A PART WITH A WHOLE

Fractions are numbers that can be used to compare a part of an object with the whole object. They can also be used to compare a part of a set of objects with the whole set of objects. We will discuss both of these uses of fractions in this section.

1. The dotted lines divide the circle at the right into four equal parts. One of the four parts is shaded. In order to compare the shaded part to the whole circle, we can use the number $\frac{1}{4}$. That is:

   $$\frac{\text{Number of shaded parts}}{\text{Total number of parts}} \longrightarrow \frac{1}{4}$$

   The dotted lines divide the square at the right into five equal parts. Two of the parts are shaded. In the space below, write the number that is used to compare the shaded parts to the whole square.

   $$\frac{\text{Number of shaded parts}}{\text{Total number of parts}} \longrightarrow \underline{}$$

2. Numbers (like $\frac{1}{4}$ and $\frac{2}{5}$) that are used to compare a part or parts to a whole object are called "fractions".

   $\frac{2}{5}$

   Write the fraction that is used to compare the shaded part or parts to the whole figure in each case at the right.

   (a) (b) (c)

   a) $\frac{2}{3}$ b) $\frac{1}{6}$ c) $\frac{7}{10}$

3. There are three circles in the set at the right. Two of them are shaded. In order to compare the number of shaded circles with the whole set of circles, we can use the fraction $\frac{2}{3}$. That is:

$$\frac{\text{Number of shaded circles}}{\text{Total number of circles}} \longrightarrow \frac{2}{3}$$

Write the fraction that is used to compare the number of shaded figures with the whole set of figures in each case below.

(a)   (b)   (c)

_____

4. We can use a fraction to compare the number of girls (or boys) in a class with the total number of students by using the formula below.

$$\frac{\text{Number of girls (or boys)}}{\text{Total number of students}}$$

Therefore: If there are 15 girls in a class of 31, the fraction is $\frac{15}{31}$.

(a) If there are 12 boys in a class of 23, the fraction is ____ .

(b) If there are 19 girls in a class of 28, the fraction is ____ .

a) $\frac{1}{3}$ b) $\frac{5}{6}$ c) $\frac{3}{8}$

a) $\frac{12}{23}$ b) $\frac{19}{28}$

5. Use a fraction to compare the given part of the set with the whole set in each case below.

(a) 9 days of a 31-day month

(b) 23 hours of a 24-hour day

(c) 3 broken boxes in a shipment of 10 boxes

(d) 9 miles traveled of a 16-mile trip

a) $\frac{9}{31}$ b) $\frac{23}{24}$ c) $\frac{3}{10}$ d) $\frac{9}{16}$

_____

1-2   THE MEANING OF THE WORDS "NUMERATOR" AND "DENOMINATOR"

Any fraction involves two numbers that are called the "numerator" and "denominator" of the fraction. We will identify the "numerator" and "denominator" of fractions in this section.

_____

6. Any fraction involves two numbers that are separated by a fraction line. There is a special name for each number. For example:

Numerator ⟶ $\frac{3}{7}$ ⟵ Denominator

The number "3" is above the fraction line. It is called the "numerator" of the fraction.

The number "7" is below the fraction line. It is called the "denominator" of the fraction.

In $\frac{5}{8}$: (a) the numerator is ____ .   (b) the denominator is ____ .

a) 5   b) 8

7. (a) In $\frac{2}{7}$, the <u>denominator</u> is ___. (b) In $\frac{1}{6}$, the <u>numerator</u> is ___.

8. (a) Write the fraction whose numerator is 4 and whose denominator is 9. ___

   (b) Write the fraction whose denominator is 16 and whose numerator is 11. ___

a) 7   b) 1

a) $\frac{4}{9}$ b) $\frac{11}{16}$

1-3 WORD NAMES FOR FRACTIONS

Just as there are word names for whole numbers, there are word names for fractions. We will discuss the word names for fractions in this section. The fractions will be limited to those whose denominators are less than 100.

9. In the table at the right, we have listed some fractions and their word names. Notice these points about the word names:

   (1) They begin with the ordinary names for the numerators.
   (2) They end with special names for the denominators.

$\frac{1}{2}$	one half	$\frac{1}{6}$	one sixth
$\frac{2}{3}$	two thirds	$\frac{3}{7}$	three sevenths
$\frac{3}{4}$	three fourths	$\frac{5}{8}$	five eighths
$\frac{4}{5}$	four fifths	$\frac{7}{9}$	seven ninths

Following the examples in the table, write the word name for each fraction below.

(a) $\frac{1}{3}$ ___   (b) $\frac{3}{5}$ ___   (c) $\frac{8}{9}$ ___

10. Write the fraction corresponding to each word name below.

    (a) one half ___   (b) four ninths ___   (c) two fifths ___

a) one third
b) three fifths
c) eight ninths

11. In the table below, we have listed some fractions with denominators from 10 to 19. Their word names are given.

a) $\frac{1}{2}$ b) $\frac{4}{9}$ c) $\frac{2}{5}$

   Note: Except for the word "twelfths", the special name for each denominator is formed by adding "<u>th</u>" or "<u>ths</u>" to its ordinary name.

   Write the fraction corresponding to each word name below.

   (a) nine fourteenths ___

   (b) eleven fifteenths ___

   (c) one sixteenth ___

$\frac{5}{10}$	five tenths	$\frac{13}{15}$	thirteen fifteenths
$\frac{1}{11}$	one eleventh	$\frac{3}{16}$	three sixteenths
$\frac{7}{12}$	seven twelfths	$\frac{11}{17}$	eleven seventeenths
$\frac{10}{13}$	ten thirteenths	$\frac{15}{18}$	fifteen eighteenths
$\frac{9}{14}$	nine fourteenths	$\frac{18}{19}$	eighteen nineteenths

a) $\frac{9}{14}$ b) $\frac{11}{15}$ c) $\frac{1}{16}$

12. In the table at the right below, we have listed some fractions whose denominators are multiples of 10. Their word names are given.

   Write the fraction corresponding to each word name below.

$\frac{7}{20}$	seven twentieths
$\frac{15}{40}$	fifteen fortieths
$\frac{37}{60}$	thirty-seven sixtieths
$\frac{79}{80}$	seventy-nine eightieths

   (a) three fortieths _____

   (b) fifty-one ninetieths _____

   (c) twenty-three seventieths _____

   a) $\frac{3}{40}$ b) $\frac{51}{90}$ c) $\frac{23}{70}$

13. Two fractions and their word names are given in the table at the right.

$\frac{8}{21}$	eight twenty-firsts
$\frac{13}{42}$	thirteen forty-seconds

   Write the fraction corresponding to each word name below.

   (a) eight twenty-sevenths _____   (b) forty sixty-fourths _____   (c) thirty-one thirty-seconds _____

   a) $\frac{8}{27}$ b) $\frac{40}{64}$ c) $\frac{31}{32}$

1-4 PAIRS OF EQUIVALENT FRACTIONS

When two fractions are equal, they are called "equivalent" fractions. We will discuss pairs of "equivalent" fractions in this section.

---

14. The two squares at the right are the same size. The <u>top</u> square is divided into two equal parts. The <u>bottom</u> square is divided into four equal parts.

   To compare the shaded part with the <u>top</u> square, we use the fraction $\frac{1}{2}$.

   To compare the shaded parts with the <u>bottom</u> square, we use the fraction $\frac{2}{4}$.

   Since the two squares are equal and the shaded parts are equal, you can see that the fractions $\frac{1}{2}$ and $\frac{2}{4}$ are equal.

   The two circles and their shaded parts at the right below are the same size.

   (a) For the <u>left</u> circle, what fraction is used to compare the shaded parts to the whole circle? _____

   (b) For the <u>right</u> circle, what fraction is used to compare the shaded parts to the whole circle? _____

   (c) Therefore, the fractions _____ and _____ are equal.

   a) $\frac{2}{3}$ b) $\frac{4}{6}$ c) $\frac{2}{3}$ and $\frac{4}{6}$

15. In the last frame, we saw that the pairs of fractions at the right are equal. When two fractions are equal, they are called "equivalent" fractions.

$\frac{1}{2} = \frac{2}{4}$ $\frac{2}{3} = \frac{4}{6}$

Each pair of rectangles below represents a pair of equivalent fractions. Write the pair of equivalent fractions in the space provided.

(a)   (b)   (c)

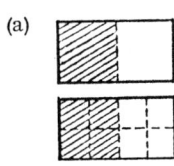

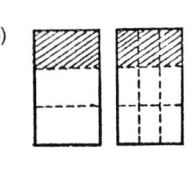

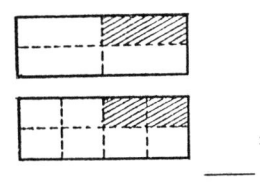

16. When two fractions are equal, we say that they are a pair of _____ fractions.

a) $\frac{1}{2} = \frac{4}{8}$ b) $\frac{1}{3} = \frac{3}{9}$ c) $\frac{1}{4} = \frac{2}{8}$

equivalent

1-5 FAMILIES OF EQUIVALENT FRACTIONS

Any fraction belongs to a family of equivalent fractions. We will discuss families of equivalent fractions in this section. The procedure used to generate a family of equivalent fractions is described.

17. Each rectangle at the right and its shaded part are the same size. Therefore, the fractions used to compare the shaded part to the whole rectangle in each case are equal or equivalent. That is:

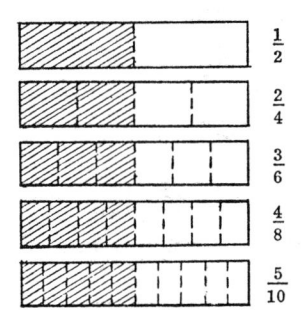

$\frac{1}{2}, \frac{2}{4}, \frac{3}{6}, \frac{4}{8}$, and $\frac{5}{10}$ are equivalent.

Since the five fractions above are equivalent or equal, we call them a "family" of equivalent fractions. We can write:

$$\frac{1}{2} = \frac{2}{4} = \frac{3}{6} = \frac{4}{8} = \frac{5}{10}$$

$\frac{1}{2}$
$\frac{2}{4}$
$\frac{3}{6}$
$\frac{4}{8}$
$\frac{5}{10}$

Using the family of equivalent fractions above, write the correct numerator in each box below.

(a) $\frac{1}{2} = \frac{\Box}{10}$ (b) $\frac{2}{4} = \frac{\Box}{8}$ (c) $\frac{3}{6} = \frac{\Box}{2}$

18. The rectangles at the right represent the following family of equivalent fractions:

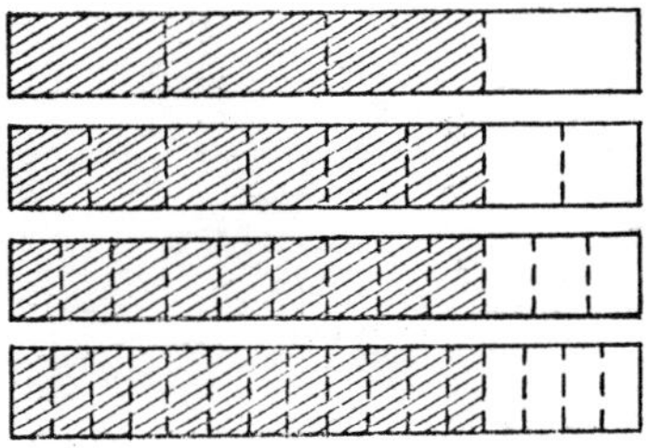

$$\frac{3}{4} = \frac{6}{8} = \frac{9}{12} = \frac{12}{16}$$

Using the family of equivalent fractions above, write the correct numerator in each box below.

$\frac{3}{4}$
$\frac{6}{8}$
$\frac{9}{12}$
$\frac{12}{16}$

a) $\frac{1}{2} = \frac{\boxed{5}}{10}$

b) $\frac{2}{4} = \frac{\boxed{4}}{8}$

c) $\frac{3}{6} = \frac{\boxed{1}}{2}$

(a) $\frac{3}{4} = \frac{\Box}{12}$ (b) $\frac{6}{8} = \frac{\Box}{16}$ (c) $\frac{12}{16} = \frac{\Box}{4}$

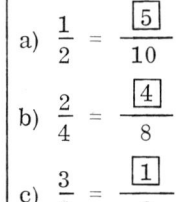

a) $\frac{\boxed{9}}{12}$ b) $\frac{\boxed{12}}{16}$ c) $\frac{\boxed{3}}{4}$

19. In an earlier frame, we saw the following family of equivalent fractions:

$$\frac{1}{2} = \frac{2}{4} = \frac{3}{6} = \frac{4}{8} = \frac{5}{10}$$

If we start with the fraction $\frac{1}{2}$, we can generate all the other fractions in the family. To do so, we simply multiply both the numerator and denominator of $\frac{1}{2}$ by the same whole number. That is:

Multiplying both by "2": $\frac{1}{2} = \frac{1 \times 2}{2 \times 2} = \frac{2}{4}$ Multiplying both by "4": $\frac{1}{2} = \frac{1 \times 4}{2 \times 4} = \frac{4}{8}$

Multiplying both by "3": $\frac{1}{2} = \frac{1 \times 3}{2 \times 3} = \frac{3}{6}$ Multiplying both by "5": $\frac{1}{2} = \frac{1 \times 5}{2 \times 5} = \frac{5}{10}$

We can use the same procedure to generate more fractions in the same family. That is:

Multiplying both by "6": $\frac{1}{2} = \frac{1 \times 6}{2 \times 6} = \frac{6}{12}$ Multiplying both by "10": $\frac{1}{2} = \frac{1 \times 10}{2 \times 10} = $ _____

20. The following family of equivalent fractions was also seen in an earlier frame:

$\frac{10}{20}$

$$\frac{3}{4} = \frac{6}{8} = \frac{9}{12} = \frac{12}{16} = \frac{15}{20}$$

The other fractions in the family can be generated by multiplying both the numerator and denominator of $\frac{3}{4}$ by the same whole number. That is:

Multiplying both by "2": $\frac{3}{4} = \frac{3 \times 2}{4 \times 2} = \frac{6}{8}$ Multiplying both by "4": $\frac{3}{4} = \frac{3 \times 4}{4 \times 4} = \frac{12}{16}$

Multiplying both by "3": $\frac{3}{4} = \frac{3 \times 3}{4 \times 3} = \frac{9}{12}$ Multiplying both by "5": $\frac{3}{4} = \frac{3 \times 5}{4 \times 5} = \frac{15}{20}$

The same procedure can be used to generate more fractions in the same family. That is:

(a) Multiplying both by "6": $\frac{3}{4} = \frac{3 \times 6}{4 \times 6} = $ _____   (b) Multiplying both by "9": $\frac{3}{4} = \frac{3 \times 9}{4 \times 9} = $ _____

21. The first three fractions in a family of equivalent fractions are: $\frac{1}{7} = \frac{2}{14} = \frac{3}{21}$ a) $\frac{18}{24}$ b) $\frac{27}{36}$

Find the next two members of the family by:

(a) multiplying both the numerator and denominator of $\frac{1}{7}$ by "4". $\frac{1}{7} = $ _____ = _____

(b) multiplying both the numerator and denominator of $\frac{1}{7}$ by "5". $\frac{1}{7} = $ _____ = _____

22. Find the next two members in each family by multiplying both the numerator and denominator of the fraction on the far left by "4" and then by "5".

(a) $\frac{3}{5} = \frac{6}{10} = \frac{9}{15} = $ _____ = _____   (b) $\frac{1}{8} = \frac{2}{16} = \frac{3}{24} = $ _____ = _____

a) $\frac{1}{7} = \frac{1 \times 4}{7 \times 4} = \frac{4}{28}$

b) $\frac{1}{7} = \frac{1 \times 5}{7 \times 5} = \frac{5}{35}$

a) $= \frac{12}{20} = \frac{15}{25}$

b) $= \frac{4}{32} = \frac{5}{40}$

1-6 RAISING FRACTIONS TO HIGHER TERMS

In this section, we will discuss the procedure for raising a fraction to higher terms. We will limit the discussion to those cases in which the denominator of the higher-terms fraction is a multiple of the denominator of the lower-terms fraction.

23. The numerator and denominator of a fraction are called the "terms" of the fraction. For example:

In $\frac{5}{8}$, 5 and 8 are called the "terms" of the fraction.

In $\frac{3}{10}$, 3 and 10 are called the _____ of the fraction.

24. Two pairs of equivalent fractions are given at the right. In each pair, the terms in the fraction on the right are larger than the corresponding terms in the fraction on the left. That is:

$\frac{1}{2} = \frac{3}{6}$ $\frac{2}{5} = \frac{4}{10}$

terms

The terms of $\frac{3}{6}$ are larger than the terms of $\frac{1}{2}$, since 3 and 6 are larger than 1 and 2.

The terms of $\frac{4}{10}$ are larger than the terms of $\frac{2}{5}$, since 4 and 10 are larger than 2 and 5.

In each pair of equivalent fractions at the right, identify the fraction whose terms are larger.

(a) $\frac{1}{3} = \frac{3}{9}$ _____   (b) $\frac{6}{8} = \frac{3}{4}$ _____

25. In any pair of equivalent fractions, we say that the fraction whose terms are larger is in higher terms. For example:

a) $\frac{3}{9}$ b) $\frac{6}{8}$

$\frac{5}{10}$ is "in higher terms" than $\frac{3}{6}$ $\frac{8}{12}$ is "in higher terms" than $\frac{6}{9}$

In each pair of equivalent fractions at the right, identify the fraction that is in higher terms.

(a) $\frac{6}{12} = \frac{2}{4}$ _____   (b) $\frac{3}{10} = \frac{6}{20}$ _____

26. We can raise any fraction to an equivalent fraction in higher terms by multiplying both its numerator and denominator by the same whole number. For example:

a) $\frac{6}{12}$ b) $\frac{6}{20}$

$\frac{4}{5} = \frac{4 \times 3}{5 \times 3} = \frac{12}{15}$ $\frac{7}{10} = \frac{7 \times 9}{10 \times 9} = \frac{63}{90}$

(a) Raise $\frac{1}{3}$ to higher terms by multiplying both terms by "2". $\frac{1}{3} =$ _____

(b) Raise $\frac{5}{7}$ to higher terms by multiplying both terms by "10". $\frac{5}{7} =$ _____

27. In each case below, we are raising a fraction to higher terms. Write the new numerator in each box.

a) $\frac{2}{6}$ b) $\frac{50}{70}$

(a) $\frac{1}{4} = \frac{1 \times 3}{4 \times 3} = \frac{\boxed{}}{12}$ (b) $\frac{5}{6} = \frac{5 \times 2}{6 \times 2} = \frac{\boxed{}}{12}$ (c) $\frac{7}{9} = \frac{7 \times 4}{9 \times 4} = \frac{\boxed{}}{36}$

a) $\frac{\boxed{3}}{12}$ b) $\frac{\boxed{10}}{12}$ c) $\frac{\boxed{28}}{36}$

28. In the problem below, you are asked to raise $\frac{1}{2}$ to an equivalent fraction in higher terms whose denominator is 16.

    $$\frac{1}{2} = \frac{\boxed{}}{16}$$

    Since the "2" was multiplied by 8 to get the new denominator "16", you must multiply the "1" by 8 to get the new numerator. Do so.

29. To find the new numerator at the right, we must multiply the "5" by some whole number. We can find that whole number by dividing 24 by 8. We get 3. Complete the problem by multiplying the "5" by 3.

    $$\frac{5}{8} = \frac{\boxed{}}{24}$$

    $\frac{1}{2} = \frac{\boxed{8}}{16}$

30. To find the new numerator at the right, we must multiply the "3" by some whole number. We can find that whole number by dividing 20 by 4. We get 5. Complete the problem by multiplying the "3" by 5.

    $$\frac{3}{4} = \frac{\boxed{}}{20}$$

    $\frac{\boxed{15}}{24}$

31. To find the new numerator at the right, we used two steps:
    (1) We divided 32 by 8 and got 4.
    (2) We multiplied 7 by 4 and got 28.

    $$\frac{7}{8} = \frac{\boxed{28}}{32}$$

    $\frac{\boxed{15}}{20}$

    Using the steps above, complete each of these: (a) $\frac{1}{4} = \frac{\boxed{}}{12}$ (b) $\frac{4}{7} = \frac{\boxed{}}{28}$

32. Complete: (a) $\frac{7}{9} = \frac{\boxed{}}{45}$ (b) $\frac{1}{10} = \frac{\boxed{}}{60}$ (c) $\frac{4}{5} = \frac{\boxed{}}{40}$ a) $\frac{\boxed{3}}{12}$ b) $\frac{\boxed{16}}{28}$

33. Complete: (a) $\frac{2}{3} = \frac{\boxed{}}{33}$ (b) $\frac{3}{8} = \frac{\boxed{}}{72}$ (c) $\frac{3}{4} = \frac{\boxed{}}{100}$ a) $\frac{\boxed{35}}{45}$ b) $\frac{\boxed{6}}{60}$ c) $\frac{\boxed{32}}{40}$

34. (a) Raise $\frac{6}{7}$ to an equivalent fraction whose denominator is 14. $\frac{6}{7} = $ _____   a) $\frac{\boxed{22}}{33}$ b) $\frac{\boxed{27}}{72}$ c) $\frac{\boxed{75}}{100}$

    (b) Raise $\frac{5}{9}$ to an equivalent fraction whose denominator is 36. $\frac{5}{9} = $ _____   a) $\frac{12}{14}$ b) $\frac{20}{36}$

SELF-TEST 1 (Frames 1-34)

1. Write the fraction that is used to compare the shaded parts to the whole rectangle.

2. Write the fraction that is used to compare the number of shaded circles with the whole set of circles.

3. Write the fraction whose denominator is 16 and whose numerator is 9.

Continued on following page.

SELF-TEST 1 (Frames 1-34) - Continued

Write the fraction corresponding to each word name below.

4. eleven twelfths _____   5. forty-three sixty-fourths _____

Complete:

6. $\dfrac{3}{4} = \dfrac{\boxed{}}{16}$ 7. $\dfrac{4}{5} = \dfrac{\boxed{}}{40}$ 8. $\dfrac{1}{8} = \dfrac{\boxed{}}{80}$ 9. $\dfrac{7}{9} = \dfrac{\boxed{}}{54}$

ANSWERS: 1. $\dfrac{5}{12}$ 3. $\dfrac{9}{16}$ 4. $\dfrac{11}{12}$ 6. $\dfrac{\boxed{12}}{16}$ 8. $\dfrac{\boxed{10}}{80}$

2. $\dfrac{6}{8}$ 5. $\dfrac{43}{64}$ 7. $\dfrac{\boxed{32}}{40}$ 9. $\dfrac{\boxed{42}}{54}$

1-7 THE CONCEPT OF DIVISIBILITY

In this section, we will discuss what is meant by the concept of divisibility. Then in the next section, we will discuss some special tests for identifying numbers that are divisible by 2, 3, and 5.

---

35. If we divide a number by 2 and the remainder is "0", we say that the number is <u>divisible</u> by 2.
    For example:

    8 <u>is</u> divisible by 2, since $8 \div 2 = 4$.

    7 <u>is not</u> divisible by 2, since $7 \div 2 = 3\ r1$.

    Which of the following are divisible by 2? _____

    (a) 3   (b) 6   (c) 10   (d) 13   (e) 28

---

36. If we divide a number by 5 and the remainder is "0", we say that the number is <u>divisible</u> by 5. For example: | (b), (c), and (e)

    15 <u>is</u> divisible by 5, since $15 \div 5 = 3$.

    22 <u>is not</u> divisible by 5, since $22 \div 5 = 4\ r2$.

    Which of the following are divisible by 5? _____

    (a) 10   (b) 13   (c) 21   (d) 30   (e) 45

---

37. In general, one whole number is divisible by another if the quotient has a "0" remainder. Which of the following are divisible by 3? _____ | (a), (d), and (e)

    (a) 6   (b) 12   (c) 14   (d) 33   (e) 35

---

38. Which of the following are divisible by 7? _____ | (a), (b), and (d)

    (a) 14   (b) 15   (c) 20   (d) 35   (e) 54

---

39. Which of the following are divisible by 8? _____ | (a) and (d)

    (a) 11   (b) 16   (c) 25   (d) 32   (e) 48

10

40. Any whole number, of course, is divisible by itself since the quotient is "1" with a "0" remainder. That is:

    6 is divisible by 6, since 6 ÷ 6 = 1   15 is divisible by 15, since 15 ÷ 15 = ____

    | (b), (d), and (e)
    | 1

41. A number can be divisible by more than one number. For example:

    6 is divisible by 2, 3, and 6.   14 is divisible by ____, ____, and ____,

    | 2, 7, and 14

42. The number "16" is divisible by which of the following numbers? _____

    (a) 10   (b) 8   (c) 6   (d) 16   (e) 4

    | (b), (d), and (e)

43. The number "36" is divisible by which of the following numbers? _____

    (a) 2   (b) 5   (c) 9   (d) 18   (e) 10

    | (a), (c), and (d)

44. The number "60" is divisible by which of the following numbers? _____

    (a) 12   (b) 10   (c) 2   (d) 4   (e) 6

    | All of them

1-8   SPECIAL TESTS FOR DIVISIBILITY BY 2, 3, AND 5

In the last section, we discussed a general test for divisibility by whole numbers. In this section, we will discuss some special tests for identifying numbers that are divisible by 2, 3, and 5.

45. The first ten numbers <u>divisible by 2</u> are: 2, 4, 6, 8, 10, 12, 14, 16, 18, 20 .

    As you can see from the list above, any number divisible by 2 has a 0, 2, 4, 6, or 8 in the "ones" place.

    Which of the following are divisible by 2? _____

    (a) 26   (b) 35   (c) 48   (d) 60   (e) 91

    | (a), (c), and (d)

46. Numbers that have a 0, 2, 4, 6 or 8 in the "ones" place are <u>divisible by 2</u>. They are called "<u>even</u>" numbers.

    For example: 2, 14, 26, 38, and 50 are called "<u>even</u>" numbers.

    Numbers that have a 1, 3, 5, 7 or 9 in the "ones" place are <u>not</u> divisible by 2. They are called "<u>odd</u>" numbers.

    For example: 1, 13, 25, 37, and 49 are called "<u>odd</u>" numbers.

    (a) 8, 52, and 98 are called _____ (even/odd) numbers.

    (b) 3, 47, and 85 are called _____ (even/odd) numbers.

    | a) even
    | b) odd

47. Only "<u>even</u>" numbers are divisible by 2. Therefore, which of the following are divisible by 2? _____

    (a) 10   (b) 37   (c) 42   (d) 76   (e) 99

    | (a), (c), and (d)

48. The first ten numbers <u>divisible by 5</u> are: 5, 10, 15, 20, 25, 30, 35, 40, 45, 50.

    As you can see from the list above, any number <u>divisible by 5</u> has a 0 or 5 in the "<u>ones</u>" place.

    Which of the following are divisible by 5? _____

        (a) 65   (b) 57   (c) 93   (d) 100   (e) 105

49. The numbers 15, 30, and 99 are divisible by 3. Notice that the <u>sum of the digits</u> in each number is divisible by 3. That is:

    In "15": $1 + 5 = 6$, and 6 is divisible by 3.
    In "30": $3 + 0 = 3$, and 3 is divisible by 3.
    In "99": $9 + 9 = 18$, and 18 is divisible by 3.

    Add the digits in each number below. In which cases is their sum divisible by 3? _____

        (a) 54   (b) 28   (c) 60   (d) 87   (e) 74

    Answer: (a), (d), and (e)

50. A number is divisible by 3 <u>only if the sum of its digits is divisible by 3</u>. That is:

    42 is divisible by 3, since $42 \div 3 = 14$
        <u>Note</u>: $4 + 2 = 6$, and 6 is divisible by 3.

    37 is not divisible by 3, since $37 \div 3 = 12 \text{ r}1$
        <u>Note</u>: $3 + 7 = 10$, and 10 is not divisible by 3.

    Let's use the test to determine whether 72 is divisible by 3.

    (a) $7 + 2 = 9$. Is 9 divisible by 3? _____

    (b) On the basis of the test, is 72 divisible by 3? _____

    (c) Divide 72 by 3 to show that it is divisible by 3. $72 \div 3 =$ _____

    Answer: (a), (c), and (d)

51. Let's use the test to determine whether 49 is divisible by 3.

    (a) $4 + 9 = 13$. Is 13 divisible by 3? _____

    (b) On the basis of the test, is 49 divisible by 3? _____

    (c) Divide 49 by 3 to show that it is <u>not</u> divisible by 3. $49 \div 3 =$ _____

    Answer: a) Yes b) Yes c) 24

52. Using the digit-sum test, identify the numbers that are divisible by 3? _____

        (a) 65   (b) 60   (c) 92   (d) 81   (e) 57

    Answer: a) No b) No c) 16 r1

53. Which of the following are divisible by 3? _____

        (a) 90   (b) 69   (c) 47   (d) 70   (e) 78

    Answer: (b), (d), and (e)

    Answer: (a), (b), and (e)

12

1-9 REDUCING FRACTIONS TO LOWER TERMS

In this section, we will discuss the procedure for reducing a fraction to lower terms. We will see that the procedure for reducing a fraction to lower terms is simply the reverse of the procedure for raising a fraction to higher terms.

54. The numerator and denominator of a fraction are called its "terms". Therefore, if two fractions are equivalent or equal, we say that the one with the larger numerator and denominator is in higher terms. That is:

    $\frac{2}{4}$ and $\frac{4}{8}$ are equal, and $\frac{4}{8}$ is in higher terms.

    Similarly, if two fractions are equivalent, we say that the one with the smaller numerator and denominator is in lower terms. That is:

    $\frac{2}{3}$ and $\frac{6}{9}$ are equal, and _____ is in lower terms.

55. A family of equivalent fractions is shown below. Notice that the terms of the fractions get larger as we move to the right in the family.

    $$\frac{2}{5} = \frac{4}{10} = \frac{6}{15} = \frac{8}{20} = \frac{10}{25}$$

    (a) Is $\frac{4}{10}$ or $\frac{6}{15}$ in higher terms? _____   (b) Is $\frac{8}{20}$ or $\frac{10}{25}$ in lower terms? _____

    $\frac{2}{3}$

56. As we have seen, any fraction can be raised to an equivalent fraction in higher terms by multiplying both its numerator and denominator by the same whole number. For example:

    $$\frac{3}{7} = \frac{3 \times 2}{7 \times 2} = \frac{6}{14} \qquad \frac{4}{6} = \frac{4 \times 5}{6 \times 5} = \frac{20}{30}$$

    To reduce a fraction to an equivalent fraction in lower terms, we simply reverse the process. That is, we divide both numerator and denominator by the same whole number. For example:

    $$\frac{6}{14} = \frac{6 \div 2}{14 \div 2} = \frac{3}{7} \qquad \frac{20}{30} = \frac{20 \div 5}{30 \div 5} = \frac{4}{6}$$

    By performing the division, reduce each of these to lower terms.   (a) $\frac{16}{20} = \frac{16 \div 4}{20 \div 4} =$ _____   (b) $\frac{12}{36} = \frac{12 \div 3}{36 \div 3} =$ _____

    a) $\frac{6}{15}$ b) $\frac{8}{20}$

57. (a) Reduce $\frac{14}{21}$ to lower terms by dividing both 14 and 21 by 7. $\frac{14}{21} =$ _____

    (b) Reduce $\frac{40}{80}$ to lower terms by dividing both 40 and 80 by 10. $\frac{40}{80} =$ _____

    a) $\frac{4}{5}$ b) $\frac{4}{12}$

58. A fraction can be reduced to lower terms only if both its numerator and denominator are divisible by the same number. For example:

    $\frac{12}{15}$ can be reduced to lower terms, since both 12 and 15 are divisible by 3.

    $\frac{15}{35}$ can be reduced to lower terms, since both 15 and 35 are divisible by _____.

    a) $\frac{2}{3}$ b) $\frac{4}{8}$

    5

59. $\frac{12}{20}$ can be reduced to lower terms in <u>two</u> different ways, since both 12 and 20 are divisible by 2 and 4.

    (a) Reduce it to lower terms by dividing each term by 2. $\frac{12}{20} =$ _____

    (b) Reduce it to lower terms by dividing each term by 4. $\frac{12}{20} =$ _____

---

60. $\frac{10}{30}$ can be reduced to lower terms in <u>three</u> different ways, since both 10 and 30 are divisible by 2, 5, and 10.   a) $\frac{6}{10}$ b) $\frac{3}{5}$

    (a) Reduce it to lower terms by dividing each term by 2. $\frac{10}{30} =$ _____

    (b) Reduce it to lower terms by dividing each term by 10. $\frac{10}{30} =$ _____

---

61. $\frac{24}{48}$ can be reduced to lower terms in <u>seven</u> different ways, since both 24 and 48 are divisible by 2, 3, 4, 6, 8, 12, and 24.   a) $\frac{5}{15}$ b) $\frac{1}{3}$

    (a) Reduce it to lower terms by dividing each term by 4. $\frac{24}{48} =$ _____

    (b) Reduce it to lower terms by dividing each term by 12. $\frac{24}{48} =$ _____

---

62. (a) Divide both terms by 13 to reduce the fraction at the right. $\frac{26}{39} =$ _____   a) $\frac{6}{12}$ b) $\frac{2}{4}$

    (b) Divide both terms by 17 to reduce the fraction at the right. $\frac{17}{68} =$ _____   a) $\frac{2}{3}$ b) $\frac{1}{4}$

1-10   REDUCING FRACTIONS TO LOWEST TERMS

We will begin this section by defining what we mean by a "fraction in lowest terms". Then we will discuss the procedure for reducing fractions to lowest terms.

---

63. If the terms of a fraction are not divisible by the same number, the fraction cannot be reduced to lower terms. For example:

   $\frac{2}{5}$ cannot be reduced to lower terms, since 2 and 5 are not divisible by the same number.

   Which of the following fractions cannot be reduced to lower terms? _____

        (a) $\frac{1}{3}$ (b) $\frac{4}{6}$ (c) $\frac{3}{4}$ (d) $\frac{5}{8}$ (e) $\frac{9}{15}$

---

64. If a fraction cannot be reduced to lower terms, we say that it is <u>IN LOWEST TERMS</u>.   (a), (c), and (d)
Which of the following fractions are <u>in lowest terms</u>? _____

        (a) $\frac{5}{15}$ (b) $\frac{4}{5}$ (c) $\frac{4}{16}$ (d) $\frac{1}{9}$ (e) $\frac{7}{11}$

                                                                                                  (b), (d), and (e)

65. Each fraction at the right can be reduced to lowest terms by dividing both terms by 2. Do so. (a) $\frac{12}{14} =$ _____ (b) $\frac{2}{16} =$ _____

66. Each fraction at the right can be reduced to lowest terms by dividing both terms by 9. Do so. (a) $\frac{18}{27} =$ _____ (b) $\frac{72}{81} =$ _____

   a) $\frac{6}{7}$ b) $\frac{1}{8}$

67. Reduce each fraction below to lowest terms.

   (a) $\frac{6}{15} =$ _____ (b) $\frac{4}{6} =$ _____ (c) $\frac{20}{25} =$ _____

   a) $\frac{2}{3}$ b) $\frac{8}{9}$

68. Reduce each fraction below to lowest terms.

   (a) $\frac{14}{35} =$ _____ (b) $\frac{5}{45} =$ _____ (c) $\frac{33}{44} =$ _____

   a) $\frac{2}{5}$ b) $\frac{2}{3}$ c) $\frac{4}{5}$

69. If both terms of a fraction are divisible by more than one number, we can reduce it to lowest terms in one step only if we divide by the largest possible number. An example is given below.

   a) $\frac{2}{5}$ b) $\frac{1}{9}$ c) $\frac{3}{4}$

   Since 12 and 16 are divisible by 2 and 4, we can reduce $\frac{12}{16}$ to lower terms in two different ways. That is:

   $$\frac{12}{16} = \frac{12 \div 2}{16 \div 2} = \frac{6}{8} \qquad \frac{12}{16} = \frac{12 \div 4}{16 \div 4} = \frac{3}{4}$$

   Notice that we only obtained a fraction in lowest terms by dividing by 4. When dividing by 2, we obtained $\frac{6}{8}$ which can be reduced further.

   (a) To reduce $\frac{6}{18}$ to lower terms, we can divide both terms by 2, 3, or 6. In order to obtain a fraction in lowest terms, we must divide by _____.

   (b) To reduce $\frac{36}{48}$ to lower terms, we can divide both terms by 2, 3, 4, 6, or 12. In order to obtain a fraction in lowest terms, we must divide by _____.

70. (a) Reduce $\frac{18}{27}$ to lowest terms by dividing both terms by 9. $\frac{18}{27} =$ _____

   (b) Reduce $\frac{70}{100}$ to lowest terms by dividing both terms by 10. $\frac{70}{100} =$ _____

   (c) Reduce $\frac{16}{64}$ to lowest terms by dividing both terms by 16. $\frac{16}{64} =$ _____

   a) 6   b) 12

71. When both terms of a fraction are divisible by more than one number, it is sometimes difficult to identify the largest possible divisor. Therefore, we sometimes use more than one step to reduce a fraction to lowest terms.

   a) $\frac{2}{3}$ b) $\frac{7}{10}$ c) $\frac{1}{4}$

   For example, we reduced $\frac{18}{54}$ to lowest terms below by dividing first by 9 and then by 2.

   $$\frac{18}{54} = \frac{2}{6} = \frac{1}{3}$$

   Continued on following page.

71. Continued

    In each case below, we reduced the fraction to lower terms by dividing both terms by 9. Complete the reduction to lowest terms.

    (a) $\frac{36}{72} = \frac{4}{8} = $ _____
    (b) $\frac{81}{108} = \frac{9}{12} = $ _____

72. Though it is more efficient to reduce a fraction to lowest terms in one step, don't be disturbed if more than one step is needed.

    Reduce each fraction at the right to lowest terms. (a) $\frac{12}{48} = $ _____ (b) $\frac{60}{100} = $ _____

    a) $\frac{1}{2}$ b) $\frac{3}{4}$

73. Since it is sometimes difficult to identify the largest possible divisor, <u>always check to see that the reduced fraction is in lowest terms</u>. That is, check to see that it cannot be reduced further.

    Reduce each fraction at the right to lowest terms. (a) $\frac{50}{100} = $ _____ (b) $\frac{48}{60} = $ _____

    a) $\frac{1}{4}$ b) $\frac{3}{5}$

74. Reduce each fraction at the right to lowest terms. (a) $\frac{60}{72} = $ _____ (b) $\frac{54}{90} = $ _____

    a) $\frac{1}{2}$ b) $\frac{4}{5}$

75. Reduce each fraction at the right to lowest terms. (a) $\frac{75}{125} = $ _____ (b) $\frac{90}{120} = $ _____

    a) $\frac{5}{6}$ b) $\frac{3}{5}$

76. A fraction can be in lowest terms even if its terms are large. Which of the following fractions are in lowest terms? _____

    (a) $\frac{22}{40}$ (b) $\frac{25}{32}$ (c) $\frac{47}{50}$ (d) $\frac{36}{45}$

    a) $\frac{3}{5}$ b) $\frac{3}{4}$

77. If possible, reduce each fraction below to lowest terms.

    (a) $\frac{33}{99} = $ _____   (b) $\frac{18}{35} = $ _____   (c) $\frac{72}{80} = $ _____

    Only (b) and (c)

78. (a) Reduce the fraction at the right to lowest terms by dividing both terms by 13. $\frac{26}{65} = $ _____

    (b) Reduce the fraction at the right to lowest terms by dividing both terms by 15. $\frac{15}{45} = $ _____

    (c) Reduce the fraction at the right to lowest terms by dividing both terms by 17. $\frac{51}{68} = $ _____

    a) $\frac{1}{3}$ b) Not possible. It is already in lowest terms. c) $\frac{9}{10}$

    a) $\frac{2}{5}$ b) $\frac{1}{3}$ c) $\frac{3}{4}$

1-11 DETERMINING WHETHER TWO FRACTIONS ARE EQUAL OR NOT

In this section, we will briefly discuss a procedure for determining whether two fractions are equal or not.

79. A family of equivalent fractions is shown at the right. $\frac{1}{3} = \frac{2}{6} = \frac{3}{9} = \frac{4}{12} = \frac{5}{15}$

   (a) Only one fraction in the family is in lowest terms. Which one? _____

   (b) If the other fractions in the family were reduced to lowest terms, the same fraction would be obtained in each case. Identify this lowest-terms fraction. _____

80. Another family of equivalent fractions is shown at the right. $\frac{2}{5} = \frac{4}{10} = \frac{6}{15} = \frac{8}{20} = \frac{10}{25}$ | a) $\frac{1}{3}$ b) $\frac{1}{3}$

   (a) Which one fraction is in lowest terms? _____

   (b) Which one fraction would be obtained if the others were reduced to lowest terms? _____

81. The point we are trying to make is this: **If two equivalent fractions are reduced to lowest terms, the same lowest-terms fraction is obtained.** | a) $\frac{2}{5}$ b) $\frac{2}{5}$

   Let's use the fact above to determine whether $\frac{9}{12}$ and $\frac{15}{20}$ are equal or not.

   (a) Reduce $\frac{9}{12}$ to lowest terms. _____

   (b) Reduce $\frac{15}{20}$ to lowest terms. _____   (c) Are $\frac{9}{12}$ and $\frac{15}{20}$ equal? _____

82. Let's determine whether $\frac{4}{6}$ and $\frac{12}{20}$ are equal or not. | a) $\frac{3}{4}$ b) $\frac{3}{4}$ c) Yes

   (a) Reduce $\frac{4}{6}$ to lowest terms. _____

   (b) Reduce $\frac{12}{20}$ to lowest terms. _____   (c) Are $\frac{4}{6}$ and $\frac{12}{20}$ equal? _____

83. Using the same steps, determine whether each pair of fractions below is equal or not. | a) $\frac{2}{3}$ b) $\frac{3}{5}$ c) No

   (a) $\frac{5}{10}$ and $\frac{20}{40}$ (b) $\frac{14}{21}$ and $\frac{15}{25}$ (c) $\frac{6}{10}$ and $\frac{21}{35}$

84. To determine whether $\frac{1}{5}$ and $\frac{4}{20}$ are equal or not, we only have to reduce $\frac{4}{20}$ to lowest terms since $\frac{1}{5}$ is already in lowest terms. | a) Equal, since both reduce to $\frac{1}{2}$
   b) Not equal
   c) Equal, since both reduce to $\frac{3}{5}$

   (a) Does $\frac{4}{20}$ reduce to $\frac{1}{5}$? _____

   (b) Are $\frac{1}{5}$ and $\frac{4}{20}$ equal? _____

| a) Yes   b) Yes

85. In each pair of fractions below, one is already in lowest terms. Reduce the other to lowest terms to determine whether they are equal or not.

(a) $\frac{18}{24}$ and $\frac{3}{4}$ (b) $\frac{7}{16}$ and $\frac{14}{32}$ (c) $\frac{8}{16}$ and $\frac{3}{8}$

| a) Equal   b) Equal   c) Not equal |

86. Which pairs of fractions below are equal? _____

(a) $\frac{8}{28}$ and $\frac{6}{18}$ (b) $\frac{70}{100}$ and $\frac{14}{20}$ (c) $\frac{1}{8}$ and $\frac{6}{32}$

| Only (b) |

SELF-TEST 2 (Frames 35-86)

1. Which of the following are divisible by 3?

   (a) 32   (b) 56   (c) 72   (d) 90

2. Reduce to lower terms by dividing each term by 12: $\frac{48}{72}$ = _____

Reduce each fraction to lowest terms.

3. $\frac{24}{32}$ = _____

4. $\frac{16}{48}$ = _____

5. $\frac{60}{75}$ = _____

6. $\frac{30}{100}$ = _____

7. Which of the following are in lowest terms?

   (a) $\frac{7}{17}$ (b) $\frac{21}{36}$ (c) $\frac{44}{62}$ (d) $\frac{23}{37}$

8. Which of the following are equal?

   (a) $\frac{18}{24}$ (b) $\frac{20}{30}$ (c) $\frac{14}{21}$ (d) $\frac{16}{36}$

ANSWERS:

1. (c), (d)   2. $\frac{4}{6}$ 3. $\frac{3}{4}$ 4. $\frac{1}{3}$ 5. $\frac{4}{5}$ 6. $\frac{3}{10}$ 7. (a), (d)   8. (b), (c)

Unit 2 THE MEANING AND EQUIVALENT FORMS OF IMPROPER FRACTIONS AND MIXED NUMBERS

Improper fractions are fractions in which the numerator is equal to or larger than the denominator. Mixed numbers are numbers that have a whole-number part and a fraction part. We will discuss the meaning and equivalent forms of improper fractions and mixed numbers in this section. The relationship between improper fractions and mixed numbers is discussed. The procedure for locating fractions and mixed numbers on the number line is shown. The relationship between fractions and division is introduced.

2-1 FRACTIONS THAT EQUAL THE NUMBER "1"

In this section, we will discuss fractions in which the numerator equals the denominator. We will show that all fractions of this type equal the number "1".

1. Next to each rectangle at the right, we have written the fraction used to compare the number of shaded parts to the total number of parts. Notice that the numerator and denominator of the bottom fraction are both "4", since:

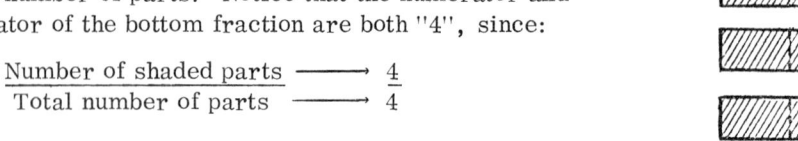

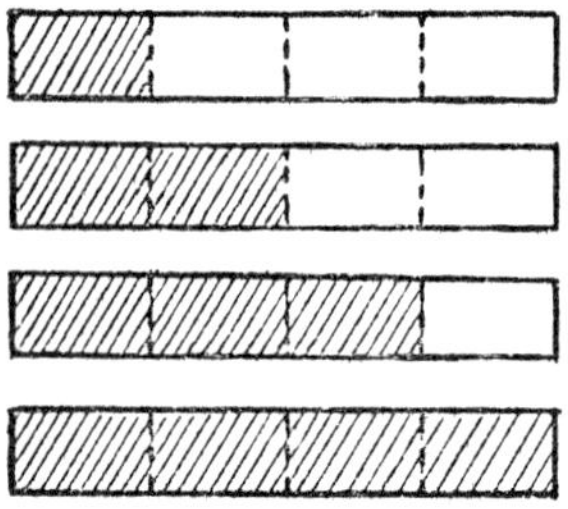

    $$\frac{\text{Number of shaded parts}}{\text{Total number of parts}} \longrightarrow \frac{4}{4}$$

    Following the pattern above, write the fraction for each bottom rectangle below.

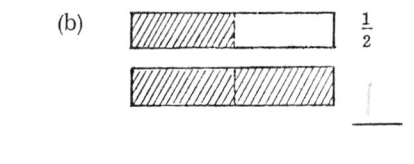

    (a) 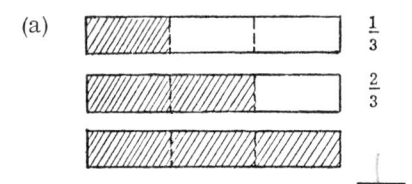 $\frac{1}{3}$, $\frac{2}{3}$, ___   (b) $\frac{1}{2}$, ___

2. We use the fraction $\frac{3}{3}$ for the circle at the right. Since all of the parts are shaded, this fraction really stands for the whole circle. Therefore, it equals the number "1". That is: $\frac{3}{3} = 1$

    We use the fraction $\frac{5}{5}$ for the circle at the right. Since all of the parts are shaded, this fraction stands for the whole circle. Therefore:

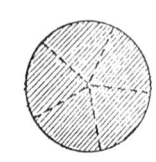

    $\frac{5}{5}$ equals what whole number? _____

    a) $\frac{3}{3}$ b) $\frac{2}{2}$

3. As you can see from the figures at the right, any fraction in which the numerator equals the denominator stands for the whole figure. Therefore, all fractions of this type equal the number "1".

   Each fraction below equals what whole number? 1

   (a) $\frac{6}{6}$ (b) $\frac{9}{9}$ (c) $\frac{10}{10}$ (d) $\frac{25}{25}$ (e) $\frac{57}{57}$

   | 1 |
   | 2/2 |
   | 3/3 |
   | 4/4 |

   | 1 |

4. The word names for fractions in which the numerator and denominator are equal follow the usual pattern. That is:

   $\frac{2}{2}$ is called "two halves". $\frac{3}{3}$ is called "three thirds". $\frac{10}{10}$ is called " _ten tenths_ ".

   | ten tenths |

2-2   FRACTIONS IN WHICH THE NUMERATOR IS LARGER THAN THE DENOMINATOR

In this section, we will show the meaning of fractions in which the numerator is larger than the denominator.

5. Each circle at the right is divided into "thirds". There are "3 thirds" shaded in one circle and "1 third" shaded in the other. That is, a total of "4 thirds" are shaded. To represent "4 thirds", we use the fraction $\frac{4}{3}$.

   Each circle at the right is divided into "fourths". A total of "7 fourths" are shaded. What fraction do we use to represent "7 fourths"? $\frac{7}{4}$

   | $\frac{7}{4}$ |

6. A total of "5 halves" are shaded at the right. To represent "5 halves", we use the fraction $\frac{5}{2}$.

   A total of "13 sixths" are shaded at the right. What fraction do we use to represent "13 sixths"? $\frac{13}{6}$

   | $\frac{13}{6}$ |

7. As we saw in the last few frames, we can obtain fractions in which the numerator is larger than the denominator. The usual method is used to name fractions of that type. For example:

   fifteen sevenths = $\frac{15}{7}$ (a) twenty-one fifths = $\frac{21}{5}$ (b) forty elevenths = $\frac{4}{11}$

   | a) $\frac{21}{5}$ b) $\frac{40}{11}$ |

20

2-3 THE MEANING OF PROPER AND IMPROPER FRACTIONS

In this section, we will define what is meant by "proper" and "improper" fractions.

8. Any fraction in which the numerator is <u>smaller than</u> the denominator is called a "<u>proper</u>" fraction. Which of the following are "<u>proper</u>" fractions? _____

   (a) $\frac{7}{8}$ (b) $\frac{8}{7}$ (c) $\frac{2}{13}$ (d) $\frac{13}{2}$ (e) $\frac{10}{11}$

   | (a), (c), and (e) |

9. Any fraction in which the numerator is <u>equal to</u> or <u>larger than</u> the denominator is called an "improper" fraction. Which of the following are "<u>improper</u>" fractions? _____

   (a) $\frac{7}{4}$ (b) $\frac{6}{7}$ (c) $\frac{7}{7}$ (d) $\frac{8}{7}$ (e) $\frac{5}{8}$

   | (a), (c), and (d) |

10. Next to each fraction below, write either "proper" or "improper".

    (a) $\frac{3}{3}$ imp   (b) $\frac{14}{15}$ p   (c) $\frac{15}{14}$ imp

    | a) improper
    | b) proper
    | c) improper

2-4 RAISING IMPROPER FRACTIONS TO HIGHER TERMS

In this section, we will discuss the procedure for raising improper fractions to higher terms.

11. The four pairs of circles at the right represent the following family of equivalent improper fractions.

    $$\frac{3}{2} = \frac{6}{4} = \frac{9}{6} = \frac{12}{8}$$

    We can generate the three larger fractions by multiplying both terms of $\frac{3}{2}$ by 2, 3, and 4. That is:

    $\frac{3}{2} = \frac{3 \times 2}{2 \times 2} = \frac{6}{4}$ $\frac{3}{2} = \frac{3 \times 3}{2 \times 3} = \frac{9}{6}$ $\frac{3}{2} = \frac{3 \times 4}{2 \times 4} = \frac{12}{8}$

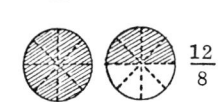

    $\frac{3}{2}$

    $\frac{6}{4}$

    $\frac{9}{6}$

    $\frac{12}{8}$

    Generate two more numbers of the same family by multiplying both terms of $\frac{3}{2}$ by 5 and 10 below.

    (a) $\frac{3}{2} = \frac{3 \times 5}{2 \times 5} = \frac{15}{10}$ (b) $\frac{3}{2} = \frac{3 \times 10}{2 \times 10} = \frac{30}{20}$

12. Any improper fraction belongs to a family of equivalent fractions. We can generate members of the family by multiplying both terms of one member by whole numbers like 2, 3, 4, and so on.

    | a) $\frac{15}{10}$ b) $\frac{30}{20}$

    Generate the next two members of the family below by multiplying both terms of $\frac{4}{3}$ by 4 and 5.

    $$\frac{4}{3} = \frac{8}{6} = \frac{12}{9} = \frac{16}{12} = \frac{20}{15}$$

    | $= \frac{16}{12} = \frac{20}{15}$

13. In the problem below, we are asked to raise $\frac{5}{2}$ to a higher-term fraction whose denominator is 8.

$$\frac{5}{2} = \frac{\boxed{20}}{8}$$

Since the "2" had to be multiplied by 4 to get the new denominator "8", we must multiply the "5" by 4 to get the new numerator. Do so.

14. To find the new numerator at the right, we must multiply the "8" by some whole number. We find that whole number by dividing 20 by 5. We get 4. Complete the problem by multiplying the "8" by 4.

$$\frac{8}{5} = \frac{\boxed{32}}{20}$$

$$\frac{\boxed{20}}{8}$$

15. To find the new numerator at the right, we used two steps:

    (1) We divided 30 by 5 and got 6.
    (2) We multiplied 2 by 6 and got 12.

$$\frac{2}{5} = \frac{\boxed{12}}{30}$$

$$\frac{\boxed{32}}{20}$$

Using the steps above, complete: (a) $\frac{3}{2} = \frac{\boxed{15}}{10}$ (b) $\frac{5}{4} = \frac{\boxed{30}}{24}$

a) $\frac{\boxed{15}}{10}$ b) $\frac{\boxed{30}}{24}$

16. Complete: (a) $\frac{11}{4} = \frac{\boxed{22}}{8}$ (b) $\frac{21}{8} = \frac{\boxed{84}}{32}$

a) $\frac{\boxed{22}}{8}$ b) $\frac{\boxed{84}}{32}$

2-5 REDUCING IMPROPER FRACTIONS TO LOWEST TERMS

In this section, we will discuss the procedure for reducing improper fractions to lower and lowest terms. We will also discuss the procedure for determining whether two improper fractions are equal or not.

17. A family of equivalent improper fractions is shown at the right.

$$\frac{4}{3} = \frac{8}{6} = \frac{12}{9} = \frac{16}{12} = \frac{20}{15} = \frac{24}{18}$$

(a) Of the two fractions $\frac{12}{9}$ or $\frac{16}{12}$, which one is in lower terms? $\frac{12}{9}$

(b) Which fraction in the family is in lowest terms? $\frac{4}{3}$

18. If both terms of an improper fraction are divisible by the same number, it can be reduced to lower terms by dividing both terms by that number. For example:

$$\frac{21}{14} = \frac{21 \div 7}{14 \div 7} = \frac{3}{2} \qquad \frac{36}{8} = \frac{36 \div 2}{8 \div 2} = \frac{18}{4} \qquad \frac{24}{9} = \frac{24 \div 3}{9 \div 3} = \frac{8}{3}$$

a) $\frac{12}{9}$ b) $\frac{4}{3}$

19. An improper fraction is in lowest terms only if its numerator and denominator are not divisible by the same whole number. Which of the following fractions are in lowest terms? _____

(a) $\frac{11}{4}$ (b) $\frac{8}{5}$ (c) $\frac{25}{10}$ (d) $\frac{42}{24}$ (e) $\frac{37}{12}$

$\frac{8}{3}$

20. Reduce each fraction to lowest terms. (a) $\frac{25}{10} = \frac{5}{2}$ (b) $\frac{21}{15} = \frac{7}{5}$

(a), (b), and (e)

a) $\frac{5}{2}$ b) $\frac{7}{5}$

21. All of the fractions at the right are equivalent or equal. If each were reduced to lowest terms, what lowest-terms fraction would be obtained? _____

$$\frac{6}{4} = \frac{9}{6} = \frac{12}{8} = \frac{15}{10} = \frac{18}{12}$$

$\frac{3}{2}$

22. A fraction can be reduced to lowest terms <u>in one step</u> only if we divide both terms by the largest possible number.

   (a) Reduce $\frac{40}{16}$ to lowest terms in one step by dividing both terms by 8. $\frac{40}{16} =$ _____

   (b) Reduce $\frac{84}{36}$ to lowest terms in one step by dividing both terms by 12. $\frac{84}{36} =$ _____

   a) $\frac{5}{2}$ b) $\frac{7}{3}$

23. Since it is not always easy to identify the largest possible divisor, it sometimes takes more than one step to reduce a fraction to lowest terms. For example, we reduced the fraction at the left below to lowest terms in two steps by dividing first by 8 and then by 2. Reduce the other two fractions to lowest terms.

   $\frac{48}{32} = \frac{6}{4} = \frac{3}{2}$ (a) $\frac{36}{24} =$ _____   (b) $\frac{80}{48} =$ _____

   a) $\frac{3}{2}$ b) $\frac{5}{3}$

24. Since it is not always easy to reduce a fraction to lowest terms in one step, <u>always check the new fraction to see that it is in lowest terms</u>.

   Reduce each fraction to lowest terms.   (a) $\frac{60}{24} =$ _____   (b) $\frac{84}{56} =$ _____

   a) $\frac{5}{2}$ b) $\frac{3}{2}$

25. Two improper fractions <u>are equal if they both reduce to the same lowest-terms fraction</u>.

   For example: $\frac{15}{10}$ and $\frac{21}{14}$ are equal, since both reduce to $\frac{3}{2}$.

   $\frac{14}{12}$ and $\frac{21}{18}$ are equal, since both reduce to _____ .

   $\frac{7}{6}$

26. Which of the following pairs of improper fractions are equal? _____

   (a) $\frac{15}{9}$ and $\frac{25}{15}$ (b) $\frac{14}{10}$ and $\frac{27}{24}$ (c) $\frac{3}{2}$ and $\frac{12}{8}$

   Both (a) and (c)

2-6   FRACTIONS THAT EQUAL WHOLE NUMBERS LARGER THAN "1"

In this section, we will discuss improper fractions that equal whole numbers larger than "1". In fractions of this type, the numerator is always a multiple of the denominator.

---

27. A total of "6 halves" are shaded at the right below. "6 halves" equals $\frac{6}{2}$. Since all of the parts in 3 circles are shaded, the fraction $\frac{6}{2}$ really equals the whole number "3". That is: $\frac{6}{2} = 3$

Continued on following page.

27. Continued

    A total of "8 fourths" are shaded at the right below. "8 fourths" equals $\frac{8}{4}$.

    Since all of the parts in 2 circles are shaded, the fraction $\frac{8}{4}$ really equals the whole number "2". That is: $\frac{8}{4} = $ _____

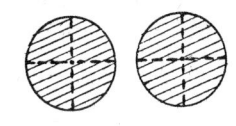

---

28. The circles at the right can be represented by the fraction $\frac{10}{5}$ or by the whole number "2". Therefore: $\frac{10}{5} = 2$

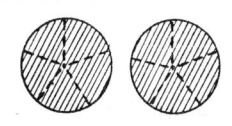

    | $\frac{8}{4} = 2$ |

    The circles at the right can be represented by the fraction $\frac{12}{3}$ or by the whole number "4". Therefore: $\frac{12}{3} = $ _____

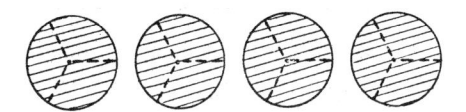

---

29. In the last few frames we saw the facts below. They show that some fractions equal whole numbers larger than "1".

    | $\frac{12}{3} = 4$ |

    $$\frac{6}{2} = 3 \qquad \frac{8}{4} = 2 \qquad \frac{10}{5} = 2 \qquad \frac{12}{3} = 4$$

    A fraction equals a whole number larger than "1" <u>only if its numerator is divisible by its denominator</u>. We can convert such a fraction to a whole number by simply performing the division. For example:

    $$\frac{15}{3} = 15 \div 3 = 5 \qquad \frac{32}{4} = 32 \div 4 = 8 \qquad \frac{42}{7} = 42 \div 7 = \underline{\quad}$$

---

30. A fraction equals a whole number <u>only if its numerator is divisible by its denominator</u>. Which fractions below equal a whole number? _____

    | 6 |

    (a) $\frac{17}{4}$ (b) $\frac{20}{2}$ (c) $\frac{70}{10}$ (d) $\frac{26}{6}$ (e) $\frac{63}{7}$

---

31. Sometimes it is difficult to see that the numerator of a fraction is divisible by the denominator. This fact only becomes apparent when we begin to reduce the fraction to lowest terms. An example is given at the left below. Convert the other two fractions to whole numbers.

    | (b), (c), and (e) |

    $$\frac{42}{14} = \frac{21}{7} = 3 \qquad (a) \ \frac{36}{18} = \underline{\quad} \qquad (b) \ \frac{96}{12} = \underline{\quad}$$

---

32. In an earlier section, we showed this fact: <u>If the numerator of a fraction equals the denominator, the fraction equals the number "1"</u>. Such a fraction can also be converted to "1" by performing a division. For example:

    | a) 2   b) 8 |

    $$\frac{3}{3} = 3 \div 3 = 1 \qquad \frac{10}{10} = 10 \div 10 = 1 \qquad \frac{59}{59} = 59 \div 59 = \underline{\quad}$$

    | 1 |

2-7 FRACTIONS WHOSE DENOMINATORS ARE "1"

In this section, we will discuss fractions whose denominators are "1". We will show that any whole number is equal to a fraction of that type.

33. Any fraction whose denominator is "1" can also be converted to a whole number by performing a division. The whole number, of course, equals the <u>numerator</u> of the fraction. For example:

$$\frac{7}{1} = 7 \div 1 = 7 \qquad \frac{15}{1} = 15 \div 1 = 15 \qquad \frac{62}{1} = 62 \div 1 = \underline{}$$

34. Any whole number can be converted to a fraction whose denominator is "1". The <u>numerator</u> of the fraction, of course, equals the whole number. For example:

$$5 = \frac{5}{1} \qquad 13 = \frac{13}{1} \qquad 20 = \frac{20}{1} \qquad 99 = \underline{}$$

| 62 |

35. Convert each fraction to a whole number and each whole number to a fraction whose denominator is "1".

(a) $\frac{9}{1} = \underline{}$ (b) $2 = \underline{}$ (c) $\frac{35}{1} = \underline{}$ (d) $81 = \underline{}$

| $\frac{99}{1}$ |

36. If we divide "1" by "1", the quotient is "1". That is, $1 \div 1 = 1$.

Therefore, the fraction $\frac{1}{1}$ equals what whole number? $\underline{}$

| a) 9 b) $\frac{2}{1}$ c) 35 d) $\frac{81}{1}$ |

37. Two families of equivalent fractions are shown at the right. Each family is equal to the whole number at the left.

$$1 = \frac{1}{1} = \frac{2}{2} = \frac{3}{3} = \frac{4}{4} = \frac{5}{5}$$

The lowest-terms fraction in the top family is $\frac{1}{1}$.

$$4 = \frac{4}{1} = \frac{8}{2} = \frac{12}{3} = \frac{16}{4} = \frac{20}{5}$$

The lowest-terms fraction in the bottom family is $\underline{}$.

| 1 |

| $\frac{4}{1}$ |

2-8 CONVERTING WHOLE NUMBERS TO FRACTIONS

We have seen that some improper fractions can be converted to whole numbers. In this section, we will show the procedure for converting whole numbers to improper fractions with denominators larger than "1".

38. Any whole number can be converted to an improper fraction in two steps. They are:

(1) Convert the whole number to a fraction whose denominator is "1".
(2) Then multiply both terms of that fraction by the same whole number.

As an example, we have converted "4" to a fraction below. Both terms of $\frac{4}{1}$ were multiplied by 6.

$$4 = \frac{4}{1} = \frac{4 \times 6}{1 \times 6} = \frac{24}{6}$$

Continued on following page.

38. Continued

    Complete each of the following conversions of a whole number to an improper fraction.

    (a) $2 = \frac{2}{1} = \frac{2 \times 7}{1 \times 7} = \underline{}$ (b) $5 = \frac{5}{1} = \frac{5 \times 3}{1 \times 3} = \underline{}$

39. By multiplying $\frac{3}{1}$ by a series of whole numbers, we can generate a family of improper fractions that equal "3". For example:

    $3 = \frac{3}{1} = \frac{3 \times 2}{1 \times 2} = \frac{6}{2}$ $3 = \frac{3}{1} = \frac{3 \times 3}{1 \times 3} = \frac{9}{3}$ $3 = \frac{3}{1} = \frac{3 \times 4}{1 \times 4} = \frac{12}{4}$

    Complete the conversions below to obtain two more members of the same family.

    (a) $3 = \frac{3}{1} = \frac{3 \times 8}{1 \times 8} = \underline{}$ (b) $3 = \frac{3}{1} = \frac{3 \times 10}{1 \times 10} = \underline{}$

    a) $\frac{14}{7}$ b) $\frac{15}{3}$

40. By multiplying both terms of $\frac{5}{1}$ by 2, 3, and 4 below, we have generated some members of the family of fractions that equal "5". Generate the next two members by multiplying both terms of $\frac{5}{1}$ by 5 and by 6.

    $5 = \frac{5}{1} = \frac{10}{2} = \frac{15}{3} = \frac{20}{4} = \underline{} = \underline{}$

    a) $\frac{24}{8}$ b) $\frac{30}{10}$

41. Generate the next two members of the family at the right. $10 = \frac{10}{1} = \frac{20}{2} = \frac{30}{3} = \underline{} = \underline{}$

    $= \frac{25}{5} = \frac{30}{6}$

42. In the problem below, you are asked to convert the whole number "2" to an improper fraction whose denominator is 6. To do so, it is helpful to begin by converting the whole number to a fraction whose denominator is "1".

    $2 = \frac{2}{1} = \frac{\boxed{}}{6}$

    Since the denominator "1" had to be multiplied by 6 to get the new denominator "6", you must multiply the numerator "2" by 6 to get the new numerator. Do so.

    $= \frac{40}{4} = \frac{50}{5}$

43. Following the procedure in the last frame, complete each of the following.

    (a) $3 = \frac{3}{1} = \frac{\boxed{}}{4}$ (b) $7 = \frac{7}{1} = \frac{\boxed{}}{8}$ (c) $10 = \frac{10}{1} = \frac{\boxed{}}{5}$

    $2 = \frac{2}{1} = \frac{\boxed{12}}{6}$

44. Complete: (a) $4 = \frac{4}{1} = \frac{\boxed{}}{10}$ (b) $6 = \frac{6}{1} = \frac{\boxed{}}{16}$

    a) $\frac{\boxed{12}}{4}$ b) $\frac{\boxed{56}}{8}$ c) $\frac{\boxed{50}}{5}$

45. When the number "1" is converted to an improper fraction, the numerator and denominator of the fraction are equal. That is:

    $1 = \frac{2}{2}$ $1 = \frac{15}{15}$ (a) $1 = \frac{\boxed{}}{7}$ (b) $1 = \frac{\boxed{}}{50}$

    a) $\frac{\boxed{40}}{10}$ b) $\frac{\boxed{96}}{16}$

    a) $\frac{\boxed{7}}{7}$ b) $\frac{\boxed{50}}{50}$

46. Complete: (a) $2 = \dfrac{\Box}{8}$ (b) $1 = \dfrac{\Box}{5}$ (c) $9 = \dfrac{\Box}{3}$

47. Complete: (a) $10 = \dfrac{\Box}{2}$ (b) $2 = \dfrac{\Box}{32}$ (c) $1 = \dfrac{\Box}{87}$

a) $\dfrac{16}{8}$	b) $\dfrac{5}{5}$	c) $\dfrac{27}{3}$
a) $\dfrac{20}{2}$	b) $\dfrac{64}{32}$	c) $\dfrac{87}{87}$

2-9 FRACTIONS WHOSE NUMERATORS ARE "0"

In this section, we will discuss fractions whose numerators are "0". We will show that all fractions of this type equal the number "0".

48. Next to each rectangle at the right, we have written the fraction used to compare the number of shaded parts to the total number of parts. Notice that the numerator of the bottom fraction is "0", since:

$$\dfrac{\text{Number of shaded parts}}{\text{Total number of parts}} \longrightarrow \dfrac{0}{3}$$

None of the parts are shaded in the rectangles below. Write the fraction used to compare the number of shaded parts with the total number of parts in each case.

(a) ▭   (b) ▭

a) $\dfrac{0}{2}$ b) $\dfrac{0}{6}$

49. We use the fraction $\dfrac{0}{3}$ for the circle at the right. Since none of the parts are shaded, this fraction really stands for the number "0". That is: $\dfrac{0}{3} = 0$.

We use the fraction $\dfrac{0}{5}$ for the circle at the right. Since none of the parts are shaded, $\dfrac{0}{5}$ equals what number? _____

0

50. Any fraction whose numerator is "0" equals the number "0". Therefore, all of the fractions at the right equal what number? _____ $\dfrac{0}{7}$ $\dfrac{0}{15}$ $\dfrac{0}{40}$

0

2-10   FRACTIONS ON THE NUMBER LINE

In this section, we will show that fractions can also be located on the number line. Just like whole numbers, they are located on the number line according to their size.

51. The rectangles at the right represent the fractions $\frac{2}{5}$ and $\frac{3}{5}$. As you can see from the figures, $\frac{3}{5}$ is the larger of the two fractions since it represents more shaded parts. That is: $\frac{3}{5}$ is larger than $\frac{2}{5}$.

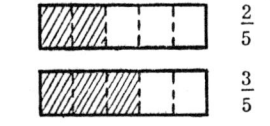

   Using the fractions provided, complete each statement below.

   (a) $\frac{1}{3}$, $\frac{2}{3}$ is larger than ____ ____

   (b) $\frac{5}{8}$, $\frac{3}{8}$ is larger than ____ ____

52. If two fractions have the same or "like" denominators, <u>the one with the larger numerator is larger</u>. Encircle the larger fraction in each pair.

   (a) $\frac{7}{10}$ and $\frac{6}{10}$ (b) $\frac{11}{8}$ and $\frac{12}{8}$ (c) $\frac{0}{5}$ and $\frac{1}{5}$

   a) $\frac{2}{3}$ is larger than $\frac{1}{3}$
   b) $\frac{5}{8}$ is larger than $\frac{3}{8}$

53. Just as we can count with whole numbers, we can count with fractions. For example, we have counted by "thirds" below.

   $\frac{0}{3}, \frac{1}{3}, \frac{2}{3}, \frac{3}{3}, \frac{4}{3}, \frac{5}{3}, \frac{6}{3}, \frac{7}{3}, \frac{8}{3}, \frac{9}{3}, \frac{10}{3}$

   List the next three fractions in the counting by "eighths" below.

   $\frac{0}{8}, \frac{1}{8}, \frac{2}{8}, \frac{3}{8}, \frac{4}{8}, \frac{5}{8}, \frac{6}{8}, \frac{7}{8},$ ____, ____, ____

   a) $\frac{7}{10}$ b) $\frac{12}{8}$ c) $\frac{1}{5}$

54. Fractions can be located on the number line. For example, we have located the "fourths" fractions on the number line below.

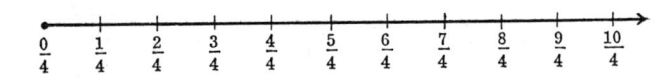

   $\frac{0}{4} \; \frac{1}{4} \; \frac{2}{4} \; \frac{3}{4} \; \frac{4}{4} \; \frac{5}{4} \; \frac{6}{4} \; \frac{7}{4} \; \frac{8}{4} \; \frac{9}{4} \; \frac{10}{4}$

   What "halves" fraction would be located at point A on the number line below?

   $\frac{0}{2} \; \frac{1}{2} \; \frac{2}{2} \; \frac{3}{2} \; \frac{4}{2} \;$ A

   $\frac{8}{8}, \frac{9}{8}, \frac{10}{8}$

55. When fractions are located on the number line, we frequently use whole numbers for the fractions equivalent to them. For example, we have used 0, 1, and 2 instead of $\frac{0}{3}$, $\frac{3}{3}$, and $\frac{6}{3}$ on the "thirds" number line below.

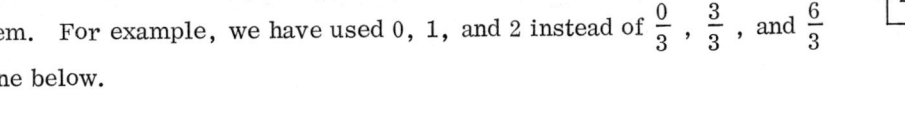

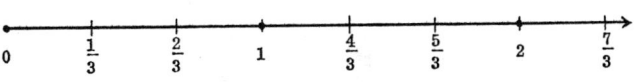

   $0 \; \frac{1}{3} \; \frac{2}{3} \; 1 \; \frac{4}{3} \; \frac{5}{3} \; 2 \; \frac{7}{3}$

   $\frac{5}{2}$

Continued on following page.

55. Continued

What whole number would be located at point A on the "sixths" number line below? _____

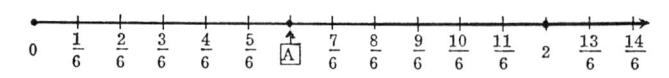

[number line showing 0, 1/6, 2/6, 3/6, 4/6, 5/6, A, 7/6, 8/6, 9/6, 10/6, 11/6, 2, 13/6, 14/6]

1

SELF-TEST 3 (Frames 1-55)

Write each word name as a fraction.

1. twenty-one halves = _____

2. ten fifths = _____

3. Which of the following are "improper" fractions?

   (a) $\frac{7}{7}$ (b) $\frac{1}{9}$ (c) $\frac{5}{8}$ (d) $\frac{17}{16}$

Complete:

4. $\frac{7}{2} = \frac{\boxed{}}{12}$ 5. $\frac{13}{5} = \frac{\boxed{}}{25}$

Reduce each fraction to lowest terms.

6. $\frac{28}{12} = $ _____   7. $\frac{45}{30} = $ _____

Convert each fraction to a whole number.

8. $\frac{60}{12} = $ _____   9. $\frac{8}{1} = $ _____

Convert each whole number to a fraction whose denominator is "1".

10. $16 = $ _____   11. $1 = $ _____

Find the numerator of each of the following fractions.

12. $8 = \frac{\boxed{}}{7}$ 13. $1 = \frac{\boxed{}}{25}$ 14. $20 = \frac{\boxed{}}{5}$ 15. $0 = \frac{\boxed{}}{16}$

ANSWERS:

1. $\frac{21}{2}$ 4. $\frac{\boxed{42}}{12}$ 6. $\frac{7}{3}$ 9. 8   12. $\frac{\boxed{56}}{7}$ 14. $\frac{\boxed{100}}{5}$

2. $\frac{10}{5}$ 5. $\frac{\boxed{65}}{25}$ 7. $\frac{3}{2}$ 10. $\frac{16}{1}$ 13. $\frac{\boxed{25}}{25}$ 15. $\frac{\boxed{0}}{16}$

3. (a), (d)   8. 5   11. $\frac{1}{1}$

2-11   ADDITION OF "LIKE" FRACTIONS

Fractions that have the same denominators are called "like" fractions. We will briefly discuss the addition of "like" fractions in this section.

56. An addition of two fractions with "like" denominators can be shown on the number line. As an example, we have added $\frac{3}{7}$ and $\frac{2}{7}$ on the number line at the right. Note the following:

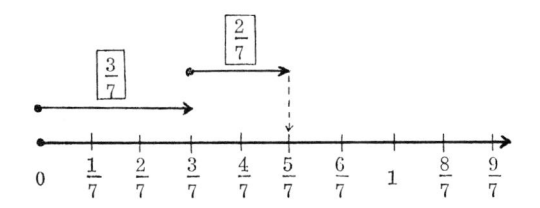

    (1) The first arrow represents $\frac{3}{7}$.

    The second arrow represents $\frac{2}{7}$.

    (2) Reading down from the tip of the second arrow, we get the answer $\frac{5}{7}$.

    Therefore: $\frac{3}{7} + \frac{2}{7} = $ _____

57. The following addition is shown on the number line at the right.

    $\frac{2}{4} + \frac{3}{4} = \frac{5}{4}$

    Write the addition shown on the number line at the right.

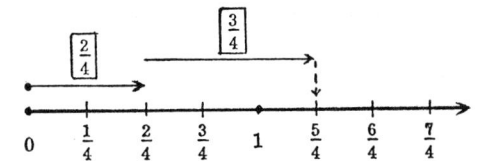

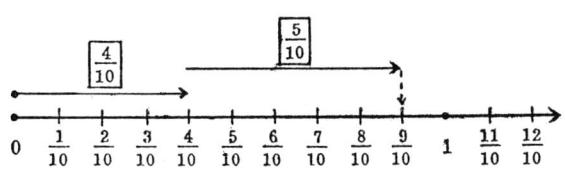

    _____ + _____ = _____

    $\boxed{\frac{5}{7}}$

58. The following three additions were performed on the number line in the last two frames.

    $\frac{3}{7} + \frac{2}{7} = \frac{5}{7}$ $\frac{2}{4} + \frac{3}{4} = \frac{5}{4}$ $\frac{4}{10} + \frac{5}{10} = \frac{9}{10}$

    $\boxed{\frac{4}{10} + \frac{5}{10} = \frac{9}{10}}$

    From the additions above you can see this fact: <u>To add two fractions with "like" denominators, we simply add their numerators and keep the same denominator.</u> Using this fact, complete each addition below.

    (a) $\frac{5}{7} + \frac{1}{7} = \frac{5+1}{7} = $ _____   (b) $\frac{11}{12} + \frac{8}{12} = \frac{11+8}{12} = $ _____

59. When two fractions are added, the answer is called the "sum". Find each sum below.   a) $\frac{6}{7}$ b) $\frac{19}{12}$

    (a) $\frac{7}{3} + \frac{7}{3} = $ _____   (b) $\frac{10}{9} + \frac{12}{9} = $ _____   (c) $\frac{1}{11} + \frac{1}{11} = $ _____

60. When adding two fractions, <u>we always reduce the sum to lowest terms if possible</u>. For example: a) $\frac{14}{3}$ b) $\frac{22}{9}$ c) $\frac{2}{11}$

    $\frac{5}{8} + \frac{1}{8} = \frac{6}{8} = \frac{3}{4}$ $\frac{7}{6} + \frac{3}{6} = \frac{10}{6} = $ _____

61. When adding two fractions, <u>we always convert the sum to a whole number if possible</u>. For example:

    $\frac{4}{7} + \frac{3}{7} = \frac{7}{7} = 1$ $\frac{11}{8} + \frac{5}{8} = \frac{16}{8} = $ _____

    $\boxed{\frac{5}{3}}$

    $\boxed{2}$

29

30

62. Perform each addition below. If possible, reduce the sum to lowest terms or convert it to a whole number.

(a) $\frac{4}{4} + \frac{6}{4}$ = _____ (b) $\frac{11}{16} + \frac{5}{16}$ = _____ (c) $\frac{17}{2} + \frac{3}{2}$ = _____

a) $\frac{5}{2}$ (from $\frac{10}{4}$) b) 1 (from $\frac{16}{16}$) c) 10 (from $\frac{20}{2}$)

2-12 THE MEANING OF MIXED NUMBERS

In this section, we will discuss the meaning of mixed numbers. We will see that any mixed number really stands for an addition of a whole number and a fraction.

63. We saw earlier that the figure at the right can be represented by the fraction $\frac{7}{3}$. Since the figure contains "2 whole shaded circles" plus "a circle with $\frac{1}{3}$ shaded", we can also represent it with the addition "$2 + \frac{1}{3}$". Therefore: $\frac{7}{3} = 2 + \frac{1}{3}$

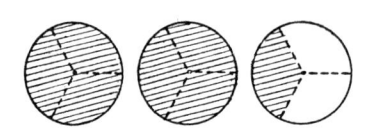

The figure at the right can be represented by the fraction $\frac{7}{4}$. Since "1 whole circle is shaded" plus "$\frac{3}{4}$ of the other circle", we can also represent it with the addition "$1 + \frac{3}{4}$". Therefore: $\frac{7}{4}$ = ____ + ____

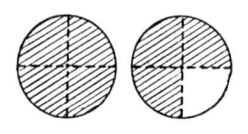

64. The figure at the right can be represented by the fraction $\frac{17}{5}$.

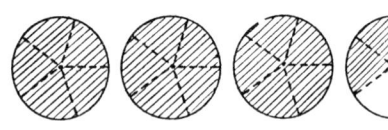

(a) What addition of a whole number and a fraction represents the same figure? ____ + ____

(b) Therefore: $\frac{17}{5}$ = ____ + ____

$\frac{7}{4} = 1 + \frac{3}{4}$

65. In the last few frames, we have seen the facts below. They show that some improper fractions are equal to an addition of a whole number and a fraction.

$\frac{7}{4} = 1 + \frac{3}{4}$ $\frac{7}{3} = 2 + \frac{1}{3}$ $\frac{17}{5} = 3 + \frac{2}{5}$

a) $3 + \frac{2}{5}$

b) $\frac{17}{5} = 3 + \frac{2}{5}$

Each addition of a whole number and a fraction above is called a "<u>mixed number</u>". It is called a "mixed" number since it contains a whole-number part and a fraction part. Though a "mixed number" is really an addition of a whole number and a fraction, the whole number and the fraction are usually written next to each other without the addition symbol. That is:

$1 + \frac{3}{4}$ is written $1\frac{3}{4}$ $2 + \frac{1}{3}$ is written $2\frac{1}{3}$ $5 + \frac{4}{9}$ is written ____

$5\frac{4}{9}$

66. Though the whole-number part and the fraction part of a mixed number are ordinarily written next to each other, remember that a mixed number stands for an addition. That is:

$1\frac{7}{8}$ means: $1 + \frac{7}{8}$ $5\frac{3}{16}$ means: $5 + \frac{3}{16}$ $9\frac{1}{4}$ means: _____ + _____

67. Convert each mixed number to an addition and each addition to a mixed number.

(a) $3\frac{1}{2}$ = ____ + ____   (b) $6 + \frac{5}{8}$ = ____   (c) $20\frac{2}{3}$ = ____ + ____

$9 + \frac{1}{4}$

68. Some mixed numbers and their word names are given in the table below. Notice that the word names contain the name of the whole number and the name of the fraction connected by the word "and".

a) $3 + \frac{1}{2}$ b) $6\frac{5}{8}$ c) $20 + \frac{2}{3}$

Write the mixed number corresponding to each word name below.

(a) three and two thirds _____

(b) five and nine tenths _____

(c) one and thirteen sixteenths _____

$1\frac{3}{4}$	one and three fourths
$2\frac{1}{8}$	two and one eighth
$10\frac{7}{12}$	ten and seven twelfths

a) $3\frac{2}{3}$

b) $5\frac{9}{10}$

c) $1\frac{13}{16}$

2-13 CONVERTING MIXED NUMBERS TO IMPROPER FRACTIONS

Any mixed number can be converted to an equivalent improper fraction. We will show both the long method and the short method for such conversions in this section.

69. Any mixed number can be converted to an improper fraction in three steps. As an example, we have converted $2\frac{3}{5}$ to an improper fraction below. Notice the steps:

(1) We wrote $2\frac{3}{5}$ as an addition.

(2) We substituted $\frac{10}{5}$ for 2.

(3) We added the two "like" fractions to get $\frac{13}{5}$.

$$2\frac{3}{5} = 2 + \frac{3}{5}$$
$$= \frac{10}{5} + \frac{3}{5} = \frac{13}{5}$$

Therefore, $2\frac{3}{5}$ is equivalent or equal to what improper fraction? _____

70. We have converted $1\frac{5}{6}$ to an improper fraction at the right. The arrow shows that we substituted $\frac{6}{6}$ for "1".

Therefore, $1\frac{5}{6}$ is equivalent to what improper fraction? _____

$$1\frac{5}{6} = 1 + \frac{5}{6}$$
$$= \frac{6}{6} + \frac{5}{6} = \frac{11}{6}$$

$\frac{13}{5}$

$\frac{11}{6}$

71. A method for converting a mixed number to an improper fraction was shown in the last two frames. In that method, the key step is converting an addition of a whole number and a fraction into an addition of two "like" fractions. For example:

$$2 + \frac{3}{5} \text{ was converted to } \frac{10}{5} + \frac{3}{5} \quad \left(\underline{\text{Note}}: \frac{10}{5} \text{ has the same denominator as } \frac{3}{5}.\right)$$

$$1 + \frac{5}{6} \text{ was converted to } \frac{6}{6} + \frac{5}{6} \quad \left(\underline{\text{Note}}: \frac{6}{6} \text{ has the same denominator as } \frac{5}{6}.\right)$$

By substituting a fraction for the whole number, convert each addition at the right into an addition of two "like" fractions.

(a) $3 + \frac{1}{2}$ (b) $1 + \frac{3}{8}$ (c) $2 + \frac{7}{12}$

_____ $+ \frac{1}{2}$ _____ $+ \frac{3}{8}$ _____ $+ \frac{7}{12}$

---

72. Complete the following conversions of a mixed number to an improper fraction.

a) $\frac{6}{2} + \frac{1}{2}$ b) $\frac{8}{8} + \frac{3}{8}$ c) $\frac{24}{12} + \frac{7}{12}$

(a) $4\frac{1}{2} = 4 + \frac{1}{2} = \underline{} + \frac{1}{2} = \underline{}$

(b) $1\frac{9}{16} = 1 + \frac{9}{16} = \underline{} + \frac{9}{16} = \underline{}$

---

73. We performed the following conversion earlier: $2\frac{3}{5} = \frac{13}{5}$

a) $\frac{8}{2} + \frac{1}{2} = \frac{9}{2}$

b) $\frac{16}{16} + \frac{9}{16} = \frac{25}{16}$

There is a shorter method that can be used for the same conversion. The steps in the shorter method are shown in the box below.

$$2\frac{3}{5} = \boxed{\frac{(5 \times 2) + 3}{5} = \frac{10 + 3}{5}} = \frac{13}{5}$$

Notice what we did to get from the mixed number $2\frac{3}{5}$ to the improper fraction $\frac{13}{5}$.

(1) <u>To get the numerator "13"</u>, we multiplied the whole number "2" by the denominator "5" and got 10. Then we added the numerator "3" to get 13.

(2) <u>To get the denominator "5"</u>, we simply used the original denominator.

We performed the following conversion earlier: $4\frac{1}{2} = \frac{9}{2}$. Use the shorter method to perform the same conversion at the right.

$$4\frac{1}{2} = \frac{(2 \times 4) + 1}{2} = \frac{\underline{} + \underline{}}{2} = \underline{}$$

---

74. We performed the following conversion earlier: $1\frac{5}{6} = \frac{11}{6}$. Use the shorter method to perform the same conversion below.

$$1\frac{5}{6} = \frac{(6 \times 1) + 5}{6} = \frac{\underline{} + \underline{}}{6} = \underline{}$$

$\frac{8+1}{2} = \frac{9}{2}$

---

75. Use the shorter method to perform each conversion at the right.

(a) $2\frac{6}{7} = \frac{(7 \times 2) + 6}{7} = \frac{\underline{} + \underline{}}{7} = \underline{}$

(b) $1\frac{3}{8} = \frac{(8 \times 1) + 3}{8} = \frac{\underline{} + \underline{}}{8} = \underline{}$

$\frac{6+5}{6} = \frac{11}{6}$

---

a) $\frac{14 + 6}{7} = \frac{20}{7}$ b) $\frac{8 + 3}{8} = \frac{11}{8}$

76. When using the shortcut, the multiplication can be done mentally. An example is shown at the right.

$$2\frac{7}{10} = \frac{20 + 7}{10} = \frac{27}{10}$$

Note: To get the "20" in the numerator of $\frac{20+7}{10}$, we multiplied 2 by 10 <u>mentally</u>.

Using the mental shortcut described above, complete: (a) $3\frac{4}{5} = \frac{+}{5} = \underline{}$ (b) $1\frac{2}{9} = \frac{+}{9} = \underline{}$

a) $\frac{15+4}{5} = \frac{19}{5}$ b) $\frac{9+2}{9} = \frac{11}{9}$

77. Convert each mixed number to a fraction.

(a) $2\frac{1}{8} = \underline{}$ (b) $1\frac{3}{4} = \underline{}$ (c) $9\frac{1}{2} = \underline{}$

a) $\frac{17}{8}$ b) $\frac{7}{4}$ c) $\frac{19}{2}$

78. Convert each of these to a fraction.   (a) $1\frac{7}{12} = \underline{}$ (b) $5\frac{9}{10} = \underline{}$

a) $\frac{19}{12}$ b) $\frac{59}{10}$

79. Convert each of these to a fraction.   (a) $12\frac{1}{3} = \underline{}$ (b) $2\frac{3}{16} = \underline{}$

a) $\frac{37}{3}$ b) $\frac{35}{16}$

2-14   CONVERTING IMPROPER FRACTIONS TO MIXED NUMBERS

Any improper fraction in which the numerator is not divisible by the denominator can be converted to an equivalent mixed number. We will show both the long and short methods for such conversions in this section.

80. If the numerator of an improper fraction is divisible by the denominator, we can convert the fraction to a whole number. For example:

$\frac{12}{4} = 3$ $\frac{7}{7} = 1$ (a) $\frac{40}{8} = \underline{}$ (b) $\frac{13}{13} = \underline{}$

a) 5   b) 1

81. If the numerator of an improper fraction is <u>not</u> divisible by the denominator, we can convert the fraction to a mixed number. An example is given below. The steps are described.

(1) We wrote $\frac{7}{2}$ as the addition $\frac{6}{2} + \frac{1}{2}$.

(2) We converted $\frac{6}{2}$ to the whole number "3".

(3) We wrote $3 + \frac{1}{2}$ as $3\frac{1}{2}$.

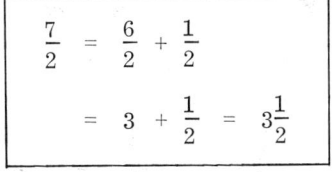

$$\frac{7}{2} = \frac{6}{2} + \frac{1}{2}$$
$$= 3 + \frac{1}{2} = 3\frac{1}{2}$$

Therefore, the improper fraction $\frac{7}{2}$ is equivalent to the mixed number $\underline{}$.

$3\frac{1}{2}$

34

82. Complete the following conversions to mixed numbers.

(a) $\frac{8}{3} = \frac{6}{3} + \frac{2}{3}$
$= 2 + \frac{2}{3} =$ _____

(b) $\frac{9}{5} = \frac{5}{5} + \frac{4}{5}$
$= 1 + \frac{4}{5} =$ _____

83. Complete the following conversions to mixed numbers.

(a) $\frac{23}{5} = \frac{20}{5} + \frac{3}{5}$
$=$ _____ $+ \frac{3}{5} =$ _____

(b) $\frac{37}{10} = \frac{30}{10} + \frac{7}{10}$
$=$ _____ $+ \frac{7}{10} =$ _____

a) $2\frac{2}{3}$ b) $1\frac{4}{5}$

84. Complete the following conversions to mixed numbers.

a) $4 + \frac{3}{5} = 4\frac{3}{5}$ b) $3 + \frac{7}{10} = 3\frac{7}{10}$

(a) $\frac{10}{3} = \frac{9}{3} + \frac{1}{3} =$ _____ $+ \frac{1}{3} =$ _____

(b) $\frac{17}{12} = \frac{12}{12} + \frac{5}{12} =$ _____ $+ \frac{5}{12} =$ _____

85. We performed the following conversion earlier: $\frac{23}{5} = 4\frac{3}{5}$. There is a shorter method that can be used to perform the same conversion. This shorter method involves division. The steps are shown below.

$$\frac{23}{5} = 5\overline{)23}^{\,4\ r3} = 4\frac{3}{5}$$

a) $3 + \frac{1}{3} = 3\frac{1}{3}$

b) $1 + \frac{5}{12} = 1\frac{5}{12}$

Notice these points about the shorter "division" method.

(1) We divided 23 by 5 and got 4 r3.

(2) The whole-number part of the mixed number is "4". It is the same as the whole-number part of the quotient.

(3) The fraction part of the mixed number is $\frac{3}{5}$. Its <u>numerator</u> "3" is the remainder in the quotient. Its <u>denominator</u> "5" is the divisor in the division.

We performed the following conversion earlier: $\frac{10}{3} = 3\frac{1}{3}$.

Complete the shorter "division" method for the same conversion at the right.

$\frac{10}{3} = 3\overline{)10}^{\,3\ r1} =$ _____

$3\frac{1}{3}$

86. When using the "division" method, remember how we obtain the fraction part of the mixed number.

The <u>remainder</u> is the <u>numerator</u> of the fraction.
The <u>divisor</u> is the <u>denominator</u> of the fraction.

Complete each conversion to a mixed number at the right.

(a) $\frac{23}{6} = 6\overline{)23}^{\,3\ r5} =$ _____

(b) $\frac{19}{16} = 16\overline{)19}^{\,1\ r3} =$ _____

87. Use the shorter "division" method below for each conversion to a mixed number.

(a) $\frac{45}{7} = 7\overline{)45} =$ _____

(b) $\frac{39}{8} = 8\overline{)39} =$ _____

(c) $\frac{47}{32} = 32\overline{)47} =$ _____

a) $3\frac{5}{6}$ b) $1\frac{3}{16}$

a) $7\overline{)45}^{\,6\ r3} = 6\frac{3}{7}$ b) $8\overline{)39}^{\,4\ r7} = 4\frac{7}{8}$ c) $32\overline{)47}^{\,1\ r15} = 1\frac{15}{32}$

88. Convert each improper fraction to a mixed number.

   (a) $\frac{17}{2}$ = _____   (b) $\frac{20}{11}$ = _____   (c) $\frac{93}{5}$ = _____

   | a) $8\frac{1}{2}$ b) $1\frac{9}{11}$ c) $18\frac{3}{5}$

89. When converting an improper fraction to a mixed number, we always reduce the fraction part of the mixed number to lowest terms. For example:

   $$\frac{12}{8} = 1\frac{4}{8} = 1\frac{1}{2} \qquad \frac{28}{6} = 4\frac{4}{6} = 4\frac{2}{3}$$

   Convert each improper fraction below to a mixed number in lowest terms.

   (a) $\frac{10}{4}$ = _____   (b) $\frac{34}{8}$ = _____   (c) $\frac{44}{12}$ = _____

90. Convert each mixed number to a fraction and each fraction to a mixed number.

   | a) $2\frac{1}{2}$ b) $4\frac{1}{4}$ c) $3\frac{2}{3}$

   (a) $1\frac{3}{8}$ = _____   (c) $2\frac{7}{32}$ = _____

   (b) $\frac{20}{8}$ = _____   (d) $\frac{104}{32}$ = _____

   | a) $\frac{11}{8}$ c) $\frac{71}{32}$
   | b) $2\frac{1}{2}$ d) $3\frac{1}{4}$

2-15   MIXED NUMBERS ON THE NUMBER LINE

In an earlier section, we showed how proper and improper fractions are located on the number line according to their size. In this section, we will show how mixed numbers are located on the number line according to their size.

91. Any mixed number lies between two consecutive whole numbers. For example:

   $2\frac{1}{4}$ lies between 2 and 3 $10\frac{7}{8}$ lies between _____ and _____

92. If two mixed numbers have different whole-number parts, the one with the larger whole-number part is larger. For example:

   | 10 and 11

   $3\frac{1}{8}$ is larger than $2\frac{5}{8}$, since 3 is larger than 2.

   Underline the larger mixed number in each pair: (a) $1\frac{3}{4}$ and $2\frac{1}{4}$ (b) $7\frac{1}{3}$ and $6\frac{2}{3}$

93. If two mixed numbers have the same whole-number parts, the one with the larger fraction part is larger. For example:

   | a) $2\frac{1}{4}$ b) $7\frac{1}{3}$

   $5\frac{7}{8}$ is larger than $5\frac{3}{8}$, since $\frac{7}{8}$ is larger than $\frac{3}{8}$.

   Underline the larger mixed number in each pair: (a) $1\frac{3}{4}$ and $1\frac{1}{4}$ (b) $9\frac{5}{8}$ and $9\frac{3}{8}$

94. Just as we can count with fractions, we can count with mixed numbers. For example, we have used mixed numbers to count by "thirds" below.

$$0, \frac{1}{3}, \frac{2}{3}, 1, 1\frac{1}{3}, 1\frac{2}{3}, 2, 2\frac{1}{3}, 2\frac{2}{3}, 3, 3\frac{1}{3}$$

List the next three numbers in the counting by "fifths" at the right.

$$0, \frac{1}{5}, \frac{2}{5}, \frac{3}{5}, \frac{4}{5}, 1, 1\frac{1}{5}, 1\frac{2}{5}, \_\_\_, \_\_\_, \_\_\_$$

a) $1\frac{3}{4}$ b) $9\frac{5}{8}$

---

95. Part of the "fourths" number line is shown at the right.

Since $1\frac{3}{4} = \frac{7}{4}$, $1\frac{3}{4}$ would be located at the same point as ____.

$1\frac{3}{5}, 1\frac{4}{5}, 2$

---

96. Part of the "eighths" number line is shown at the right.

$1\frac{3}{8}$ would be located at the same point as $\frac{11}{8}$. $2\frac{1}{8}$ would be located at the same point as ____.

$\frac{7}{4}$

---

97. In the last two frames, we recorded improper fractions on the number line. Ordinarily, however, proper fractions are recorded between all whole numbers as we have done for part of the "sixths" number line below.

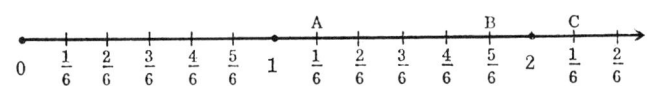

When proper fractions are recorded between all whole numbers, the points on the line are read as mixed numbers. For example:

Point A is read $1\frac{1}{6}$. The fraction $\frac{7}{6}$ would be located at point A.

Point B is read $1\frac{5}{6}$. The fraction $\frac{11}{6}$ would be located at point B.

Point C is read $2\frac{1}{6}$. What improper fraction would be located at point C? ____

$\frac{17}{8}$

---

98. Part of the "halves" number line is shown at the right.

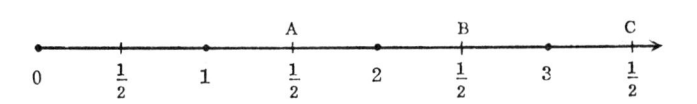

(a) Point A represents the mixed number $1\frac{1}{2}$. Point B represents the mixed number ____.

(b) $\frac{3}{2}$ would be located at point A. $\frac{7}{2}$ would be located at point ____.

$\frac{13}{6}$

---

a) $2\frac{1}{2}$ b) C

2-16 RELATING FRACTIONS AND DIVISION

Up to this point, we have seen that some fractions are related to division. In this section, we will show that any fraction is merely another way of writing a division.

99. We saw this fact earlier: <u>Any improper fraction in which the numerator is divisible by the denominator can be converted to a whole number by performing a division.</u> For example:

$$\frac{24}{8} = 24 \div 8 = 3 \qquad \frac{5}{5} = 5 \div 5 = 1 \qquad \frac{35}{7} = 35 \div 7 = \underline{}$$

100. We also saw this fact earlier: <u>Any improper fraction in which the numerator is not divisible by the denominator can be converted to a mixed number by performing a division.</u> For example:

$$\frac{11}{2} = 2\overline{)11}^{\,5\,r1} = 5\frac{1}{2} \qquad \frac{25}{7} = 7\overline{)25}^{\,3\,r4} = 3\frac{4}{7} \qquad \frac{20}{3} = 3\overline{)20} = \underline{}$$

⎤ 5

101. It should be obvious that any improper fraction is equal to a division in which the numerator is divided by the denominator. That is:

$$\frac{20}{5} = 20 \div 5 \qquad \frac{10}{10} = 10 \div 10 \qquad \frac{14}{9} = \underline{} \div \underline{}$$

⎤ $3\overline{)20}^{\,6\,r2} = 6\frac{2}{3}$

102. Write each division as an <u>improper</u> fraction. (a) $40 \div 8 = \underline{}$ (b) $49 \div 16 = \underline{}$

⎤ $14 \div 9$

103. Although we cannot make the following fact clear until decimal numbers are introduced, it is also true that <u>any proper fraction is equal to a division in which the numerator is divided by the denominator.</u> That is:

$$\frac{2}{5} = 2 \div 5 \qquad \frac{7}{10} = 7 \div 10 \qquad \frac{1}{3} = \underline{} \div \underline{}$$

⎤ a) $\frac{40}{8}$ b) $\frac{49}{16}$

104. Write each division as a <u>proper</u> fraction. (a) $3 \div 7 = \underline{}$ (b) $1 \div 4 = \underline{}$

⎤ $1 \div 3$

105. We saw this fact earlier: <u>Any fraction whose numerator is "0" is equal to the number "0".</u> This fact can be confirmed by converting the fraction to a division. That is:

$$\frac{0}{4} = 0 \div 4 = 0 \qquad \frac{0}{9} = 0 \div 9 = 0 \qquad \frac{0}{25} = 0 \div 25 = \underline{}$$

⎤ a) $\frac{3}{7}$ b) $\frac{1}{4}$

106. Even divisions containing large numbers can be converted to fractions. For example:

$$746 \div 15 = \frac{746}{15} \qquad 235 \div 1{,}847 = \frac{235}{1{,}847}$$

Convert this division to a fraction: $625{,}000 \div 4{,}500 = \underline{}$

⎤ 0

⎤ $\dfrac{625{,}000}{4{,}500}$

2-17 WRITING ANY QUOTIENT WITH A REMAINDER AS A MIXED NUMBER

When converting improper fractions to mixed numbers by performing divisions, we saw how smaller quotients with remainders can be written as mixed numbers. In this section, we will show that any quotient with a remainder can be written as a mixed number.

107. When converting improper fractions to mixed numbers by the division method, we saw that a quotient with a remainder can be written as a mixed number. For example:

$$\frac{11}{4} = 4\overline{)11}^{\,2\,r3} = 2\frac{3}{4} \qquad\qquad \frac{29}{8} = 8\overline{)29}^{\,3\,r5} = 3\frac{5}{8}$$

Even when the numbers involved are larger, a quotient with a remainder can be written as a mixed number. For example:

The quotient of $23\overline{)396}^{\,17\,r5}$ can be written $17\frac{5}{23}$. The quotient of $140\overline{)4{,}629}^{\,33\,r9}$ can be written _____ .

108. Write each quotient as a mixed number.   (a) $69\overline{)4{,}001}^{\,57\,r68}$ _____   (b) $19\overline{)20{,}379}^{\,1{,}072\,r11}$ _____

   $33\frac{9}{140}$

109. When writing a quotient with a remainder as a mixed number, we always reduce the fraction part of the mixed number to lowest terms. For example:

   a) $57\frac{68}{69}$ b) $1{,}072\frac{11}{19}$

The quotient of $64\overline{)1{,}760}^{\,27\,r32}$ is written $27\frac{32}{64}$ and reduced to $27\frac{1}{2}$.

Write each quotient below as a mixed number. Be sure to reduce the fraction part of the mixed number to lowest terms.

(a) $36\overline{)207}^{\,5\,r27}$ _____   (b) $120\overline{)65{,}680}^{\,547\,r40}$ _____

a) $5\frac{3}{4}$ b) $547\frac{1}{3}$

SELF-TEST 4 (Frames 56-109)

1. Write $9 + \frac{13}{16}$ as a mixed number.	2. Write "eleven and four fifths" as a mixed number.
Convert to an improper fraction:	Convert to a mixed number:
3. $1\frac{3}{4} =$ _____ 4. $7\frac{2}{5} =$ _____	5. $\frac{53}{6} =$ _____ 6. $\frac{28}{16} =$ _____
Convert to a whole number:	9. Write $17 \div 32$ as a fraction. _____
7. $\frac{18}{3} =$ _____ 8. $\frac{12}{12} =$ _____	10. Write $17\overline{)65}^{\,3\,r14}$ as a mixed number. _____

ANSWERS: 1. $9\frac{13}{16}$ 3. $\frac{7}{4}$ 5. $8\frac{5}{6}$ 7. 6   9. $\frac{17}{32}$

2. $11\frac{4}{5}$ 4. $\frac{37}{5}$ 6. $1\frac{3}{4}$ 8. 1   10. $3\frac{14}{17}$

Unit 3 ADDITION AND SUBTRACTION OF FRACTIONS

In this unit, we will discuss the addition and subtraction of both "like" and "unlike" fractions. We will show that the lowest common denominator for any addition or subtraction is the smallest common multiple of the denominators. A special section is devoted to a method for comparing the size of fractions.

3-1 ADDING FRACTIONS WITH "LIKE" DENOMINATORS

The procedure for adding two fractions with the same or "like" denominators was discussed in the last unit. We will briefly review the procedure in this section.

1. To add two fractions with "like" denominators, we simply add their numerators and keep the same denominator. For example:

    $$\frac{3}{7} + \frac{1}{7} = \frac{3+1}{7} = \frac{4}{7} \qquad \frac{4}{11} + \frac{5}{11} = \frac{4+5}{11} = \underline{}$$

2. When adding fractions, <u>we always reduce the sum to lowest terms if possible</u>. For example: | $\frac{9}{11}$

    $$\frac{5}{12} + \frac{5}{12} = \frac{10}{12} = \frac{5}{6} \qquad \frac{3}{6} + \frac{1}{6} = \frac{4}{6} = \underline{2/3}$$

3. When adding two fractions, the sum can be an improper fraction. That is: | $\frac{2}{3}$

    $$\frac{9}{10} + \frac{2}{10} = \frac{11}{10} \qquad \frac{5}{7} + \frac{8}{7} = \frac{13}{7} \qquad \frac{14}{3} + \frac{6}{3} = \underline{20/6}$$

4. Sometimes an improper-fraction sum can be reduced to a whole number. For example: | $\frac{20}{3}$

    $$\frac{5}{6} + \frac{1}{6} = \frac{6}{6} = 1 \qquad \frac{3}{2} + \frac{7}{2} = \frac{10}{2} = \underline{5}$$

5. When the sum is an improper fraction that does not equal a whole number, we convert it to a mixed number. For example: | 5

    $$\frac{7}{5} + \frac{1}{5} = \frac{8}{5} = 1\frac{3}{5} \qquad \frac{9}{10} + \frac{12}{10} = \frac{21}{10} = \underline{2\frac{1}{10}}$$

6. When a sum is converted to a mixed number, <u>the fraction part of the mixed number is always reduced to lowest terms if possible</u>. For example: | $2\frac{1}{10}$

    $$\frac{7}{12} + \frac{11}{12} = \frac{18}{12} = 1\frac{6}{12} = 1\frac{1}{2} \qquad \frac{5}{6} + \frac{11}{6} = \frac{16}{6} = \underline{2\frac{4}{6}\,\frac{2}{3}}$$

7. Remember that sums are always reported in lowest terms. For example:

$$\frac{7}{16} + \frac{5}{16} = \frac{12}{16} = \frac{3}{4} \qquad \frac{15}{8} + \frac{7}{8} = \frac{22}{8} = 2\frac{6}{8} = 2\frac{3}{4} \qquad \frac{17}{5} + \frac{3}{5} = \frac{20}{5} = 4$$

Report each sum below in lowest terms.

(a) $\frac{17}{6} + \frac{13}{6} =$ _____   (b) $\frac{16}{25} + \frac{4}{25} =$ _____   (c) $\frac{9}{8} + \frac{11}{8} =$ _____

| a) 5   b) $\frac{4}{5}$ c) $2\frac{1}{2}$ |

8. Report each sum below in lowest terms.

(a) $\frac{10}{16} + \frac{3}{16} =$ _13/16_

(b) $\frac{4}{11} + \frac{7}{11} =$ _11/11 = 1_   (c) $\frac{5}{4} + \frac{5}{4} =$ _10/8 1⅖_

| a) $\frac{13}{16}$ b) 1   c) $2\frac{1}{2}$ |

3-2 SUBTRACTING FRACTIONS WITH "LIKE" DENOMINATORS

In this section, we will show the procedure for subtracting fractions with the same or "like" denominators.

9. A subtraction of two fractions can be done on the number line. As an example, we have subtracted $\frac{2}{7}$ from $\frac{5}{7}$ on the number line at the right. Note the following:

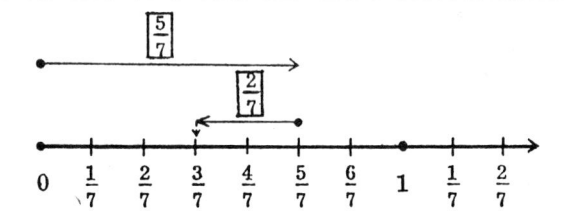

(1) The top arrow represents $\frac{5}{7}$. The bottom arrow represents $\frac{2}{7}$.

(2) Reading down from the tip of the bottom arrow, we get the answer $\frac{3}{7}$.

Therefore: $\frac{5}{7} - \frac{2}{7} =$ _3/7_

| $\frac{3}{7}$ |

10. The following subtraction is shown on the number line at the right.

$$\frac{4}{5} - \frac{3}{5} = \frac{1}{5}$$

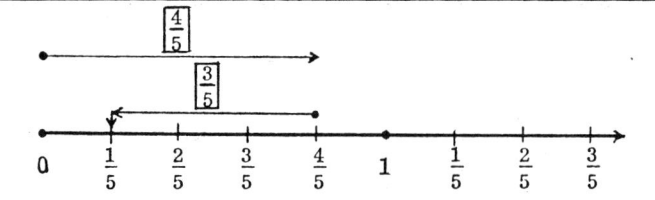

Write the subtraction shown on the number line at the right.

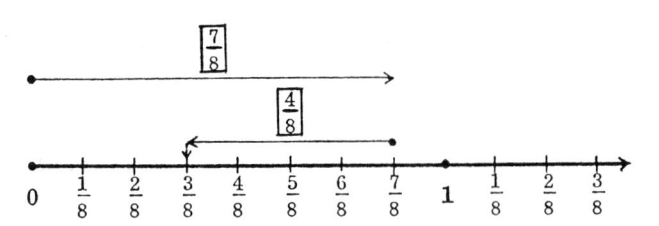

_7/8_ − _4/8_ = _3/8_

| $\frac{7}{8} - \frac{4}{8} = \frac{3}{8}$ |

11. In the last two frames, we performed the three subtractions below on the number line.

$$\frac{5}{7} - \frac{2}{7} = \frac{3}{7} \qquad \frac{4}{5} - \frac{3}{5} = \frac{1}{5} \qquad \frac{7}{8} - \frac{4}{8} = \frac{3}{8}$$

From the examples above, you can see this fact: <u>To subtract two fractions, we can subtract the second numerator from the first numerator and keep the same denominator.</u> That is:

$$\frac{5}{7} - \frac{2}{7} = \frac{5-2}{7} = \frac{3}{7} \qquad \frac{4}{5} - \frac{3}{5} = \frac{4-3}{5} = \frac{1}{5} \qquad \frac{7}{8} - \frac{4}{8} = \frac{7-4}{8} = \underline{\frac{3}{8}}$$

12. The answer in a subtraction of fractions is called the "<u>remainder</u>" or "<u>difference</u>". Find each difference below.

(a) $\frac{8}{9} - \frac{6}{9} = \underline{\frac{2}{9}}$ (b) $\frac{2}{3} - \frac{1}{3} = \underline{\frac{2}{3}}$ (c) $\frac{17}{7} - \frac{12}{7} = \underline{\frac{5}{7}}$

$\boxed{\frac{3}{8}}$

13. When subtracting fractions, <u>we always reduce the difference to lowest terms if possible.</u> For example:

$$\frac{5}{6} - \frac{2}{6} = \frac{3}{6} = \frac{1}{2} \qquad \frac{15}{16} - \frac{3}{16} = \frac{12}{16} = \underline{}$$

$\boxed{\text{a) } \frac{2}{9} \quad \text{b) } \frac{1}{3} \quad \text{c) } \frac{5}{7}}$

14. When subtracting two fractions, the difference can be an improper fraction. That is:

$$\frac{17}{12} - \frac{4}{12} = \frac{13}{12} \qquad \frac{33}{16} - \frac{6}{16} = \frac{27}{16} \qquad \frac{20}{3} - \frac{10}{3} = \underline{}$$

$\boxed{\frac{3}{4}}$

15. Sometimes an improper-fraction difference can be reduced to a whole number. For example:

$$\frac{46}{5} - \frac{1}{5} = \frac{45}{5} = 9 \qquad \frac{11}{9} - \frac{2}{9} = \frac{9}{9} = \underline{}$$

$\boxed{\frac{10}{3}}$

16. When the difference is an improper fraction that does not equal a whole number, we convert it to a mixed number. For example:

$$\frac{21}{12} - \frac{4}{12} = \frac{17}{12} = 1\frac{5}{12} \qquad \frac{45}{16} - \frac{8}{16} = \frac{37}{16} = \underline{}$$

$\boxed{1}$

17. When a difference is converted to a mixed number, <u>the fraction part of the mixed number is always reduced to lowest terms if possible.</u> For example:

$$\frac{27}{8} - \frac{7}{8} = \frac{20}{8} = 2\frac{4}{8} = 2\frac{1}{2} \qquad \frac{19}{16} - \frac{1}{16} = \frac{18}{16} = \underline{1\frac{2}{16}\ 1\frac{1}{8}}$$

$\boxed{2\frac{5}{16}}$

18. Remember that differences are always reported in lowest terms. For example:

$$\frac{19}{20} - \frac{3}{20} = \frac{16}{20} = \frac{4}{5} \qquad \frac{25}{4} - \frac{1}{4} = \frac{24}{4} = 6 \qquad \frac{19}{6} - \frac{11}{6} = \frac{8}{6} = 1\frac{2}{6} = 1\frac{1}{3}$$

$\boxed{1\frac{1}{8}}$

Report each difference below in lowest terms.

(a) $\frac{13}{7} - \frac{6}{7} = \underline{\frac{7}{7} = 1}$ (b) $\frac{15}{32} - \frac{11}{32} = \underline{\frac{4}{32}\ \frac{1}{8}}$ (c) $\frac{27}{10} - \frac{3}{10} = \underline{\frac{24}{10}\ 2\frac{4}{10}\ 2\frac{2}{5}}$

$\boxed{\text{a) } 1 \quad \text{b) } \frac{1}{8} \quad \text{c) } 2\frac{2}{5}}$

41

42

19. When we subtract a fraction from itself, the numerator of the difference is "0". Therefore, the difference is "0". For example:

$$\frac{3}{5} - \frac{3}{5} = \frac{0}{5} = 0 \qquad \frac{17}{12} - \frac{17}{12} = \frac{0}{12} = 0 \qquad \frac{42}{3} - \frac{42}{3} = \frac{0}{3} = \underline{\qquad}$$

20. In each subtraction below, the second numerator is larger than the first numerator.

$$\frac{2}{5} - \frac{4}{5} = \frac{2-4}{5} \qquad \qquad \frac{12}{7} - \frac{19}{7} = \frac{12-19}{7}$$

Since subtractions like "2 - 4" and "12 - 19" cannot be done in arithmetic, the subtractions of fractions above cannot be done in arithmetic.

Which of the following subtractions cannot be done in arithmetic? _____

(a) $\frac{7}{8} - \frac{7}{8}$ (b) $\frac{5}{9} - \frac{8}{9}$ (c) $\frac{9}{10} - \frac{8}{10}$ (d) $\frac{41}{32} - \frac{56}{32}$

| 0 |

| (b) and (d) |

3-3 THE MEANING OF MULTIPLES

In this section, we will review the meaning of the concept of "multiples" and discuss the general test for deciding whether one whole number is a multiple of another. As we will show in later sections, the "multiple" concept is needed to perform additions and subtractions of fractions with "unlike" denominators.

21. Whenever we multiply any whole number by another whole number, the product is called a "multiple" of the original number.

If we multiply 6 by 4 and get 24, we say that 24 is a multiple of 6.

If we multiply 3 by 5 and get 15, we say that 15 is a multiple of 3.

22. At the right, we have multiplied the number "4" by a series of whole numbers. All of the products are called "multiples of 4".

As you can see by examining the products, we can obtain the multiples of 4 by "counting by 4's". Using that method, complete the first nine multiples of 4 below.

4, 8, 12, 16, 20, 24, 28, 32, 36

1 x 4 = 4
2 x 4 = 8
3 x 4 = 12
4 x 4 = 16
5 x 4 = 20

| 15 |

23. To decide whether a number is a multiple of 4 or not, we divide the number by 4. The number is a multiple of 4 <u>only if it is divisible by 4</u> (that is, <u>only if the remainder is "0"</u>). For example:

84 <u>is</u> a multiple of 4, since 84 ÷ 4 = 21 .

93 <u>is not</u> a multiple of 4, since 93 ÷ 4 = 23 r1 .

Which of the following are multiples of 4? A C E

(a) 12 (b) 25 (c) 36 (d) 47 (e) 60

| 28, 32, 36 |

| (a), (c), and (e) |

24. A number is a multiple of 5 <u>only if it is divisible by 5</u>. Which of the following are multiples of 5?

    (a) 20   (b) 45   (c) 39   (d) 105   (e) 96   _____

25. A number is a multiple of 7 <u>only if it is divisible by 7</u>. Which of the following are multiples of 7? _____ | (a), (b), and (d)

    (a) 21   (b) 32   (c) 45   (d) 70   (e) 92

26. In general, one whole number is a multiple of another only if it is <u>divisible</u> by that number. | (a) and (d)

    (a) Is 18 a multiple of 3? ____   (b) Is 26 a multiple of 6? ____   (c) Is 46 a multiple of 9? ____

27. (a) Is 55 a multiple of 111? ____   (b) Is 73 a multiple of 10? ____ | a) Yes   b) No   c) No

| a) No   b) No

3-4 ADDING FRACTIONS WHEN THE LARGER DENOMINATOR IS A MULTIPLE OF THE SMALLER

Fractions can also be added when they have different or "unlike" denominators. In this section, we will discuss the procedure for adding fractions with "unlike" denominators when the larger denominator is a multiple of the smaller.

28. The fractions $\frac{1}{2}$ and $\frac{1}{4}$ have different or "unlike" denominators. We have <u>added</u> them on the "fourths" number line at the right. As you can see, their sum is $\frac{3}{4}$.

    The sum $\frac{3}{4}$ cannot be obtained by simply adding the numerators of the two original fractions and using one of the denominators. That is:

    $\frac{1}{2} + \frac{1}{4}$ does not equal $\frac{1+1}{4}$ or $\frac{2}{4}$, and $\frac{1}{2} + \frac{1}{4}$ does not equal $\frac{1+1}{2}$ or $\frac{2}{2}$

    When two fractions have "unlike" denominators, can we find their sum by simply adding their numerators and using one of the denominators? _____

29. In the last frame, we added $\frac{1}{2}$ and $\frac{1}{4}$ on the number line and got $\frac{3}{4}$ as their sum. The ordinary method for the same addition is shown at the right. Notice the steps: | No

    $\frac{1}{2} + \frac{1}{4} = \frac{2}{4} + \frac{1}{4} = \frac{3}{4}$

    (1) We substituted $\frac{2}{4}$ for $\frac{1}{2}$ to get "$\frac{2}{4} + \frac{1}{4}$" which is an addition of fractions with "like" denominators.

    (2) Then we added $\frac{2}{4}$ and $\frac{1}{4}$ in the usual way and got $\frac{3}{4}$.

    Since $\frac{2}{4} = \frac{1}{2}$: (a) Is the addition "$\frac{2}{4} + \frac{1}{4}$" equivalent to the addition "$\frac{1}{2} + \frac{1}{4}$"? _____

    (b) Is the sum for "$\frac{2}{4} + \frac{1}{4}$" equal to the sum for "$\frac{1}{2} + \frac{1}{4}$"? _____

30. We have added $\frac{1}{4}$ and $\frac{5}{8}$ on the "eighths" number line at the right. As you can see, their sum is $\frac{7}{8}$.

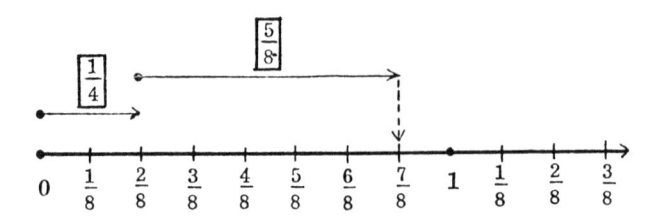

a) Yes
b) Yes

The ordinary method for the same addition is shown at the right. Notice the steps:

$$\frac{1}{4} + \frac{5}{8} = \frac{2}{8} + \frac{5}{8} = \frac{7}{8}$$

(1) We substituted $\frac{2}{8}$ for $\frac{1}{4}$ to get "$\frac{2}{8} + \frac{5}{8}$" which is an addition of "like" fractions.

(2) Then we added $\frac{2}{8}$ and $\frac{5}{8}$ in the usual way and got $\frac{7}{8}$.

Since $\frac{2}{8} = \frac{1}{4}$: (a) Is the addition "$\frac{2}{8} + \frac{5}{8}$" equivalent to the addition "$\frac{1}{4} + \frac{5}{8}$"? _____

(b) Is the sum for "$\frac{2}{8} + \frac{5}{8}$" equal to the sum for "$\frac{1}{4} + \frac{5}{8}$"? _____

31. In the last two frames, we showed the procedure for adding two "unlike" fractions <u>when the larger denominator is a multiple of the smaller</u>. That is:

a) Yes
b) Yes

In $\frac{1}{2} + \frac{1}{4}$, the "4" is a multiple of the "2". In $\frac{1}{4} + \frac{5}{8}$, the "8" is a multiple of the "4".

In which additions below is the larger denominator a multiple of the smaller? _____

(a) $\frac{3}{10} + \frac{1}{5}$ (b) $\frac{2}{3} + \frac{1}{7}$ (c) $\frac{9}{8} + \frac{11}{16}$ (d) $\frac{7}{12} + \frac{1}{8}$

32. In the addition $\frac{3}{8} + \frac{5}{16}$, 16 is a multiple of 8. To perform the addition, we use the following steps:

(a) and (c)

(1) Since $\frac{6}{16}$ is equivalent to $\frac{3}{8}$, substitute $\frac{6}{16}$ for $\frac{3}{8}$ to get "$\frac{6}{16} + \frac{5}{16}$" which is an equivalent addition of "like" fractions.

$$\frac{3}{8} + \frac{5}{16} = \frac{6}{16} + \frac{5}{16}$$

$$\frac{3}{8} + \frac{5}{16} = \frac{6}{16} + \frac{5}{16} = \frac{11}{16}$$

(2) Then perform the addition of "like" fractions in the usual way.

In the addition $\frac{1}{6} + \frac{2}{3}$, 6 is a multiple of 3. Let's perform the addition by following the steps above.

(a) Since $\frac{4}{6}$ is equivalent to $\frac{2}{3}$, substitute $\frac{4}{6}$ for $\frac{2}{3}$ to get an equivalent addition of "like" fractions.

$$\frac{1}{6} + \frac{2}{3} = \frac{1}{6} + \underline{}$$

(b) Then perform the addition of "like" fractions in the usual way.

$$\frac{1}{6} + \frac{2}{3} = \frac{1}{6} + \underline{} = \underline{}$$

a) $\frac{1}{6} + \frac{4}{6}$ b) $\frac{1}{6} + \frac{4}{6} = \frac{5}{6}$

45

33. In order to perform an addition of fractions with "unlike" denominators, we must convert it to an equivalent addition with "like" denominators. If the larger denominator is a multiple of the smaller, we can convert to an equivalent addition with "like" denominators by substituting for the fraction with the smaller denominator. For example:

For $\frac{1}{2} + \frac{3}{8}$, we substitute $\frac{4}{8}$ for $\frac{1}{2}$ and get $\frac{4}{8} + \frac{3}{8}$.

For $\frac{3}{10} + \frac{2}{5}$, we substitute $\frac{4}{10}$ for $\frac{2}{5}$ and get $\frac{3}{10} + \frac{4}{10}$.

(a) Complete the addition at the right by replacing $\frac{2}{3}$ with an equivalent fraction whose denominator is 9.

$\frac{2}{3} + \frac{1}{9} = $ _____ $+ \frac{1}{9} = $ _____

(b) Complete the addition at the right by replacing $\frac{3}{4}$ with an equivalent fraction whose denominator is 16.

$\frac{1}{16} + \frac{3}{4} = \frac{1}{16} + $ _____ $ = $ _____

34. By substituting for the fraction with the smaller denominator, complete each addition below.

(a) $\frac{5}{14} + \frac{2}{7} = \frac{5}{14} + $ _____ $ = $ _____

(b) $\frac{1}{2} + \frac{3}{8} = $ _____ $+ \frac{3}{8} = $ _____

a) $\frac{6}{9} + \frac{1}{9} = \frac{7}{9}$

b) $\frac{1}{16} + \frac{12}{16} = \frac{13}{16}$

35. Perform each addition below. Reduce each sum to lowest terms.

(a) $\frac{1}{4} + \frac{5}{12} = $ _____ $+ \frac{5}{12} = $ _____

(b) $\frac{1}{6} + \frac{1}{3} = \frac{1}{6} + $ _____ $ = $ _____

a) $\frac{5}{14} + \frac{4}{14} = \frac{9}{14}$

b) $\frac{4}{8} + \frac{3}{8} = \frac{7}{8}$

36. In each addition below, convert the sum to a mixed number with the fraction part in lowest terms.

(a) $\frac{3}{4} + \frac{1}{2} = \frac{3}{4} + $ _____ $ = $ _____

(b) $\frac{5}{6} + \frac{7}{24} = $ _____ $+ \frac{7}{24} = $ _____

a) $\frac{2}{3}$ (from $\frac{3}{12} + \frac{5}{12} = \frac{8}{12}$)

b) $\frac{1}{2}$ (from $\frac{1}{6} + \frac{2}{6} = \frac{3}{6}$)

37. Write each sum below in lowest terms.

(a) $\frac{3}{10} + \frac{13}{40} = $ _____

(b) $\frac{8}{15} + \frac{4}{5} = $ _____

a) $1\frac{1}{4}$ (from $\frac{3}{4} + \frac{2}{4} = \frac{5}{4}$)

b) $1\frac{1}{8}$ (from $\frac{20}{24} + \frac{7}{24} = \frac{27}{24}$)

a) $\frac{5}{8}$ b) $1\frac{1}{3}$

3-5 SUBTRACTING FRACTIONS WHEN THE LARGER DENOMINATOR IS A MULTIPLE OF THE SMALLER

Fractions can also be subtracted when they have "unlike" denominators. In this section, we will discuss the procedure for subtracting fractions when the larger denominator is a multiple of the smaller.

38. The fractions $\frac{1}{2}$ and $\frac{3}{4}$ have "unlike" denominators. We have subtracted $\frac{1}{2}$ from $\frac{3}{4}$ on the "fourths" number line at the right. As you can see, their difference is $\frac{1}{4}$.

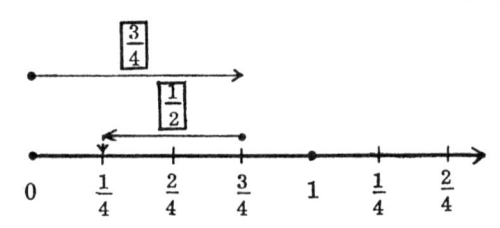

    The difference $\frac{1}{4}$ cannot be obtained by simply subtracting the numerators of the two original fractions and using one of the denominators. That is:

    $\frac{3}{4} - \frac{1}{2}$ does <u>not</u> equal $\frac{3-1}{4}$ or $\frac{2}{4}$, and $\frac{3}{4} - \frac{1}{2}$ does <u>not</u> equal $\frac{3-1}{2}$ or $\frac{2}{2}$

    When two fractions have "unlike" denominators, can we find their difference by simply subtracting the numerators and using one of the denominators? _____

39. In the last frame, we subtracted $\frac{1}{2}$ from $\frac{3}{4}$ and got $\frac{1}{4}$ as the difference. The ordinary method for the same subtraction is shown at the right. Notice that the steps are similar to those used in addition. That is:

    $\frac{3}{4} - \frac{1}{2} = \frac{3}{4} - \frac{2}{4} = \frac{1}{4}$ | No

    (1) We substituted $\frac{2}{4}$ for $\frac{1}{2}$ to get "$\frac{3}{4} - \frac{2}{4}$" which is a subtraction of "like" fractions.

    (2) Then we performed the subtraction "$\frac{3}{4} - \frac{2}{4}$" in the usual way and got $\frac{1}{4}$.

    Is the subtraction "$\frac{3}{4} - \frac{2}{4}$" equivalent to the subtraction "$\frac{3}{4} - \frac{1}{2}$"? _____

40. In the subtraction $\frac{15}{16} - \frac{5}{8}$, 16 is a multiple of 8. Perform the subtraction by following the steps below. | Yes

    (a) Substitute $\frac{10}{16}$ for $\frac{5}{8}$ to obtain an equivalent subtraction of "like" fractions.
    $\frac{15}{16} - \frac{5}{8} = \frac{15}{16} - \underline{}$

    (b) Then perform the subtraction of "like" fractions in the usual way.
    $\frac{15}{16} - \frac{5}{8} = \frac{15}{16} - \underline{} = \underline{}$

41. (a) Complete the subtraction at the right by replacing $\frac{1}{3}$ with an equivalent fraction whose denominator is 6.
    $\frac{7}{6} - \frac{1}{3} = \frac{7}{6} - \underline{} = \underline{}$

    a) $\frac{15}{16} - \frac{10}{16}$
    b) $\frac{15}{16} - \frac{10}{16} = \frac{5}{16}$

    (b) Complete the subtraction at the right by replacing $\frac{3}{4}$ with an equivalent fraction whose denominator is 8.
    $\frac{3}{4} - \frac{5}{8} = \underline{} - \frac{5}{8} = \underline{}$

42. By substituting for the fraction with the smaller denominator, complete each subtraction below.

    (a) $\dfrac{9}{14} - \dfrac{3}{7} = \dfrac{9}{14} - \underline{} = \underline{}$ (b) $\dfrac{1}{2} - \dfrac{3}{8} = \underline{} - \dfrac{3}{8} = \underline{}$

    | a) $\dfrac{7}{6} - \dfrac{2}{6} = \dfrac{5}{6}$ |
    | b) $\dfrac{6}{8} - \dfrac{5}{8} = \dfrac{1}{8}$ |

43. Do each subtraction below. Reduce each difference to lowest terms.

    (a) $\dfrac{3}{4} - \dfrac{5}{12} = \underline{}$ (b) $\dfrac{2}{3} - \dfrac{1}{6} = \underline{}$

    | a) $\dfrac{9}{14} - \dfrac{6}{14} = \dfrac{3}{14}$ |
    | b) $\dfrac{4}{8} - \dfrac{3}{8} = \dfrac{1}{8}$ |

44. In each case below, convert the difference to a mixed number. <u>Be sure</u> to <u>reduce</u> the <u>fraction part</u> to <u>lowest terms if possible</u>.

    (a) $\dfrac{9}{4} - \dfrac{1}{2} = \underline{}$ (b) $\dfrac{37}{12} - \dfrac{3}{4} = \underline{}$

    | a) $\dfrac{1}{3}$ (from $\dfrac{9}{12} - \dfrac{5}{12} = \dfrac{4}{12}$) |
    | b) $\dfrac{1}{2}$ (from $\dfrac{4}{6} - \dfrac{1}{6} = \dfrac{3}{6}$) |

45. Do each subtraction below.

    (a) $\dfrac{13}{24} - \dfrac{1}{6} = \underline{}$ (b) $\dfrac{99}{32} - \dfrac{11}{16} = \underline{}$

    | a) $1\dfrac{3}{4}$ (from $\dfrac{9}{4} - \dfrac{2}{4} = \dfrac{7}{4}$) |
    | b) $2\dfrac{1}{3}$ (from $\dfrac{37}{12} - \dfrac{9}{12} = \dfrac{28}{12}$) |

46. When the subtraction below is converted to a subtraction of "like" fractions, the second numerator is larger than the first numerator. Therefore, we cannot perform the subtraction in arithmetic.

    $$\dfrac{3}{4} - \dfrac{7}{8} = \dfrac{6}{8} - \dfrac{7}{8}$$

    Which of the following subtractions cannot be performed in arithmetic? _____

    (a) $\dfrac{2}{3} - \dfrac{5}{6}$ (b) $\dfrac{11}{6} - \dfrac{19}{12}$ (c) $\dfrac{77}{30} - \dfrac{27}{10}$

    | a) $\dfrac{3}{8}$ (from $\dfrac{9}{24}$) |
    | b) $2\dfrac{13}{32}$ (from $\dfrac{77}{32}$) |
    | (a) and (c) |

SELF-TEST 5 (Frames 1-46)

Do these additions and subtractions. Reduce answers to lowest terms.

| 1. $\dfrac{9}{20} + \dfrac{3}{20} =$ | 2. $\dfrac{37}{8} - \dfrac{7}{8} =$ | 3. $\dfrac{5}{12} + \dfrac{19}{12} =$ | 4. $\dfrac{4}{5} - \dfrac{4}{5} =$ |

5. $\dfrac{2}{3} + \dfrac{5}{6} =$ 7. $\dfrac{1}{10} + \dfrac{13}{20} =$

6. $\dfrac{1}{4} - \dfrac{1}{12} =$ 8. $\dfrac{7}{2} - \dfrac{11}{16} =$

ANSWERS: 1. $\dfrac{3}{5}$ 3. 2   5. $1\dfrac{1}{2}$ 7. $\dfrac{3}{4}$

2. $3\dfrac{3}{4}$ 4. 0   6. $\dfrac{1}{6}$ 8. $2\dfrac{13}{16}$

3-6 USING FAMILIES OF EQUIVALENT FRACTIONS FOR ADDITIONS IN WHICH THE LARGER DENOMINATOR IS NOT A MULTIPLE OF THE SMALLER

In this section, we will show how families of equivalent fractions can be used to add fractions when the larger denominator is not a multiple of the smaller. The term "common" denominator is also introduced.

47. There are two types of additions of fractions with "unlike" denominators.

   (1) Those in which the larger denominator <u>is a multiple</u> of the smaller. For example:

   $$\frac{1}{2} + \frac{3}{8} \qquad \frac{3}{10} + \frac{2}{5} \qquad \frac{5}{6} + \frac{7}{12}$$

   (2) Those in which the larger denominator <u>is not a multiple</u> of the smaller. For example:

   $$\frac{2}{3} + \frac{1}{5} \qquad \frac{1}{2} + \frac{2}{9} \qquad \frac{4}{7} + \frac{1}{3}$$

   In which additions below is the larger denominator <u>not a multiple</u> of the smaller? _____

   (a) $\frac{3}{4} + \frac{5}{12}$ (b) $\frac{2}{7} + \frac{4}{9}$ (c) $\frac{4}{5} + \frac{1}{2}$ (d) $\frac{15}{16} + \frac{5}{8}$

48. When the larger denominator <u>is a multiple</u> of the smaller, we can obtain an equivalent addition of "like" fractions <u>by making a single substitution for the fraction with the smaller denominator</u>. For example:

   | (b) and (c)

   $$\frac{1}{2} + \frac{3}{8} = \frac{4}{8} + \frac{3}{8} \qquad \text{(Note: We simply substituted } \frac{4}{8} \text{ for } \frac{1}{2}\text{)}$$

   When the larger denominator <u>is not a multiple</u> of the smaller, we cannot obtain an equivalent addition of "like" fractions <u>by making a single substitution</u>. We have to substitute for <u>both</u> fractions. An example is given below.

   To perform $\frac{2}{3} + \frac{1}{5}$, we must substitute for <u>both</u> fractions. Some members of the families of fractions equivalent to $\frac{2}{3}$ and $\frac{1}{5}$ are shown below.

   $$\left[\frac{2}{3}\right] = \frac{4}{6} = \frac{6}{9} = \frac{8}{12} = \left(\frac{10}{15}\right) = \frac{12}{18}$$

   $$\left[\frac{1}{5}\right] = \frac{2}{10} = \left(\frac{3}{15}\right) = \frac{4}{20} = \frac{5}{25} = \frac{6}{30}$$

   As you can see from the circled member in each family, both fractions are equivalent to a fraction whose denominator is 15. Therefore, we can obtain an equivalent addition of "like" fractions by substituting the two circled fractions above. That is:

   $$\frac{2}{3} + \frac{1}{5} = \frac{10}{15} + \frac{3}{15}$$

   Since $\frac{10}{15}$ equals $\frac{2}{3}$ and $\frac{3}{15}$ equals $\frac{1}{5}$, "$\frac{10}{15} + \frac{3}{15}$" is equivalent to "$\frac{2}{3} + \frac{1}{5}$".

   Therefore: $\frac{2}{3} + \frac{1}{5} = \frac{10}{15} + \frac{3}{15} =$ _____

   | $\frac{13}{15}$

49. In the addition $\frac{1}{2} + \frac{2}{5}$, the larger denominator is not a multiple of the smaller. To obtain an equivalent addition of "like" fractions, we must make a substitution for <u>both fractions</u>. Some members of the families of fractions equivalent to $\frac{1}{2}$ and $\frac{2}{5}$ are shown below.

$$\left[\frac{1}{2}\right] = \frac{2}{4} = \frac{3}{6} = \frac{4}{8} = \boxed{\frac{5}{10}} = \frac{6}{12}$$

$$\left[\frac{2}{5}\right] = \boxed{\frac{4}{10}} = \frac{6}{15} = \frac{8}{20} = \frac{10}{25} = \frac{12}{30}$$

As you can see from the circled member in each family, both fractions are equivalent to a fraction whose denominator is 10. Therefore, we can obtain an equivalent addition of "like" fractions by substituting the two circled fractions above. That is:

$$\left[\frac{1}{2}\right] + \left[\frac{2}{5}\right] = \frac{5}{10} + \frac{4}{10}$$

Since $\frac{5}{10}$ equals $\frac{1}{2}$ and $\frac{4}{10}$ equals $\frac{2}{5}$, "$\frac{5}{10} + \frac{4}{10}$" is equivalent to "$\frac{1}{2} + \frac{2}{5}$".

Therefore: $\frac{1}{2} + \frac{2}{5} = \frac{5}{10} + \frac{4}{10} = \underline{\frac{9}{10}}$

---

50. By substituting the circled members of each family at the right, complete the addition below.

$$\frac{1}{3} + \frac{4}{7} = \underline{} + \underline{} = \underline{}$$

$$\left[\frac{1}{3}\right] = \frac{2}{6} = \frac{3}{9} = \frac{4}{12} = \frac{5}{15} = \frac{6}{18} = \boxed{\frac{7}{21}}$$

$$\left[\frac{4}{7}\right] = \frac{8}{14} = \boxed{\frac{12}{21}} = \frac{16}{28} = \frac{20}{35} = \frac{24}{42} = \frac{28}{49}$$

| $\frac{9}{10}$ |

---

51. The additions performed in the last three frames are shown below. In each case, we began by obtaining an equivalent addition of "like" fractions.

| $\frac{7}{21} + \frac{12}{21} = \frac{19}{21}$ |

Addition A
$$\frac{2}{3} + \frac{1}{5} = \frac{10}{15} + \frac{3}{15}$$

Addition B
$$\frac{1}{2} + \frac{2}{5} = \frac{5}{10} + \frac{4}{10}$$

Addition C
$$\frac{1}{3} + \frac{4}{7} = \frac{7}{21} + \frac{12}{21}$$

The "like" denominators in the new additions are usually called the "<u>common</u>" denominators. That is:

In addition A, 15 is called the "common" denominator.
In addition B, 10 is called the "common" denominator.
In addition C, 21 is called the "<u>Common</u>" denominator.

---

52. As you can see from the circled members of each family at the right, we can use 20 as the "common" denominator for the addition below. Do so.

$$\frac{3}{5} + \frac{1}{4} = \underline{} + \underline{} = \underline{}$$

$$\left[\frac{3}{5}\right] = \frac{6}{10} = \frac{9}{15} = \boxed{\frac{12}{20}} = \frac{15}{25}$$

$$\left[\frac{1}{4}\right] = \frac{2}{8} = \frac{3}{12} = \frac{4}{16} = \boxed{\frac{5}{20}}$$

| common |

| $\frac{12}{20} + \frac{5}{20} = \frac{17}{20}$ |

49

3-7 THE "LOWEST COMMON DENOMINATOR" CONCEPT

In this section, we will show what is meant by a "lowest common denominator". Lowest common denominators will be used to perform some additions in which the larger denominator is not a multiple of the smaller.

53. Some members of the families of fractions equivalent to $\frac{3}{4}$ and $\frac{1}{6}$ are shown at the right. As you can see from the circled fractions, we can use either 12 or 24 as the common denominator when adding them.

$\boxed{\frac{3}{4}} = \frac{6}{8} = \boxed{\frac{9}{12}} = \frac{12}{16} = \frac{15}{20} = \boxed{\frac{18}{24}}$

$\boxed{\frac{1}{6}} = \boxed{\frac{2}{12}} = \frac{3}{18} = \boxed{\frac{4}{24}} = \frac{5}{30} = \frac{6}{36}$

(a) Use 12 as the common denominator to perform the addition. $\quad \frac{3}{4} + \frac{1}{6} = \frac{\square}{12} + \frac{\square}{12} = \underline{\qquad}$

(b) Use 24 as the common denominator to perform the addition. $\quad \frac{3}{4} + \frac{1}{6} = \frac{\square}{24} + \frac{\square}{24} = \underline{\qquad}$

(c) Was the same sum obtained in both cases? ____

54. Some members of the families of fractions equivalent to $\frac{1}{6}$ and $\frac{3}{8}$ are shown at the right. As you can see from the circled fractions, we can use either 24 or 48 as the common denominator when adding them.

$\boxed{\frac{1}{6}} = \frac{2}{12} = \frac{3}{18} = \boxed{\frac{4}{24}} = \frac{5}{30} = \frac{6}{36} = \frac{7}{42} = \boxed{\frac{8}{48}}$

$\boxed{\frac{3}{8}} = \frac{6}{16} = \boxed{\frac{9}{24}} = \frac{12}{32} = \frac{15}{40} = \boxed{\frac{18}{48}} = \frac{21}{56} = \frac{24}{64}$

a) $\frac{9}{12} + \frac{2}{12} = \frac{11}{12}$

b) $\frac{18}{24} + \frac{4}{24} = \frac{22}{24} = \frac{11}{12}$

c) Yes

(a) Use 24 as the common denominator to perform the addition. $\quad \frac{1}{6} + \frac{3}{8} = \frac{\square}{24} + \frac{\square}{24} = \underline{\qquad}$

(b) Use 48 as the common denominator to perform the addition. $\quad \frac{1}{6} + \frac{3}{8} = \frac{\square}{48} + \frac{\square}{48} = \underline{\qquad}$

(c) Was the same sum obtained in both cases? ____

55. In the last two frames, we used two different common denominators to perform the same additions.

For $\frac{3}{4} + \frac{1}{6}$, we used both 12 and 24 as the common denominators.

For $\frac{1}{6} + \frac{3}{8}$, we used both 24 and 48 as the common denominators.

a) $\frac{4}{24} + \frac{9}{24} = \frac{13}{24}$

b) $\frac{8}{48} + \frac{18}{48} = \frac{26}{48} = \frac{13}{24}$

c) Yes

When more than one common denominator can be used, <u>the smallest one is called the "lowest common denominator"</u>. That is:

For $\frac{3}{4} + \frac{1}{6}$, the "lowest common denominator" is 12.

For $\frac{1}{6} + \frac{3}{8}$, the "lowest common denominator" is ____.

56. Some members of the families of fractions equivalent to $\frac{1}{2}$ and $\frac{1}{3}$ are shown at the right. As you can see from the circled fractions, we can use either 6 or 12 as the common denominator when adding them.

$\left[\frac{1}{2}\right] = \frac{2}{4} = \left(\frac{3}{6}\right) = \frac{4}{8} = \frac{5}{10} = \left(\frac{6}{12}\right)$

$\left[\frac{1}{3}\right] = \left(\frac{2}{6}\right) = \frac{3}{9} = \left(\frac{4}{12}\right) = \frac{5}{15} = \frac{6}{18}$

(a) The lowest common denominator for $\frac{1}{2} + \frac{1}{3}$ is ____.

(b) Use the lowest common denominator to perform the addition. $\frac{1}{2} + \frac{1}{3} =$ ____ + ____ = ____

a) 6

b) $\frac{3}{6} + \frac{2}{6} = \frac{5}{6}$

57. Some members of the families of fractions equivalent to $\frac{5}{8}$ and $\frac{3}{10}$ are shown at the right. As you can see from the circled fractions, we can use either 40 or 80 as the common denominator when adding them.

$\left[\frac{5}{8}\right] = \frac{10}{16} = \frac{15}{24} = \frac{20}{32} = \left(\frac{25}{40}\right) = \frac{30}{48} = \frac{35}{56} = \frac{40}{64} = \frac{45}{72} = \left(\frac{50}{80}\right)$

$\left[\frac{3}{10}\right] = \frac{6}{20} = \frac{9}{30} = \left(\frac{12}{40}\right) = \frac{15}{50} = \frac{18}{60} = \frac{21}{70} = \left(\frac{24}{80}\right) = \frac{27}{90} = \frac{30}{100}$

(a) The lowest common denominator for $\frac{5}{8} + \frac{3}{10}$ is ____.

(b) Use the lowest common denominator to perform the addition. $\frac{5}{8} + \frac{3}{10} =$ ____ + ____ = ____

a) 40

b) $\frac{25}{40} + \frac{12}{40} = \frac{37}{40}$

58. Some members of the families of fractions equivalent to $\frac{2}{3}$ and $\frac{1}{4}$ are shown at the right. Examine the families to answer the questions below.

$\left[\frac{2}{3}\right] = \frac{4}{6} = \frac{6}{9} = \frac{8}{12} = \frac{10}{15} = \frac{12}{18} = \frac{14}{21} = \frac{16}{24}$

$\left[\frac{1}{4}\right] = \frac{2}{8} = \frac{3}{12} = \frac{4}{16} = \frac{5}{20} = \frac{6}{24} = \frac{7}{28} = \frac{8}{32}$

(a) The lowest common denominator for $\frac{2}{3} + \frac{1}{4}$ is ____.

(b) Use the lowest common denominator to perform the addition. $\frac{2}{3} + \frac{1}{4} =$ ____ + ____ = ____

a) 12

b) $\frac{8}{12} + \frac{3}{12} = \frac{11}{12}$

59. Some members of the families of fractions equivalent to $\frac{5}{6}$ and $\frac{1}{9}$ are listed at the right. Examine the families to answer the questions below.

$\left[\frac{5}{6}\right] = \frac{10}{12} = \frac{15}{18} = \frac{20}{24} = \frac{25}{30} = \frac{30}{36} = \frac{35}{42} = \frac{40}{48} = \frac{45}{54}$

$\left[\frac{1}{9}\right] = \frac{2}{18} = \frac{3}{27} = \frac{4}{36} = \frac{5}{45} = \frac{6}{54} = \frac{7}{63} = \frac{8}{72} = \frac{9}{81}$

(a) The lowest common denominator for $\frac{5}{6} + \frac{1}{9}$ is ____.

(b) Use the lowest common denominator to perform the addition. $\frac{5}{6} + \frac{1}{9} =$ ____ + ____ = ____

a) 18

b) $\frac{15}{18} + \frac{2}{18} = \frac{17}{18}$

51

52

3-8 A STRATEGY FOR IDENTIFYING LOWEST COMMON DENOMINATORS

We will begin this section by showing that the lowest common denominator is the smallest number that is a multiple of both denominators. Then we will discuss a strategy for identifying lowest common denominators.

---

60. In the last section, we used the lowest common denominator to perform various additions. For example:

　　We used "12" to add $\frac{2}{3}$ and $\frac{1}{4}$.　　　　We used "40" to add $\frac{5}{8}$ and $\frac{3}{10}$.

　　As you can see, the "lowest common denominator" is the smallest number that is a multiple of both denominators. That is:

　　12 is the smallest number that is a multiple of both 3 and 4 .

　　40 is the smallest number that is a multiple of both ___ and ___ .

---

61. Since the "lowest common denominator" is the smallest number that is a multiple of both denominators, we can find it by listing the multiples of both denominators until we find the smallest common one. An example is given below. | 8 and 10

　　To find the least common denominator for $\frac{3}{8} + \frac{5}{12}$, we have listed some multiples of 8 and 12 below.

　　　　8, 16, 24, 32, 40, 48

　　　　12, 24, 36, 48, 60, 72

(a) Identify the smallest common multiple of both 8 and 12. _____

(b) Therefore, the lowest common denominator for $\frac{3}{8} + \frac{5}{12}$ is _____.

---

62. To find the lowest common denominator for　　　　10, 20, 30, 40, 50, 60, 70　　　　a) 24

$\frac{7}{10} + \frac{5}{12}$, we have listed some multiples of　　　　12, 24, 36, 48, 60, 72, 84　　　　b) 24

10 and 12 at the right.

(a) Identify the smallest common multiple of both 10 and 12. _____

(b) Therefore, the lowest common denominator for $\frac{7}{10} + \frac{5}{12}$ is _____.

---

63. To find the lowest common denominator for $\frac{2}{3} + \frac{4}{7}$,　　3, 6, 9, 12, 15, 18, 21, 24　　a) 60

we have listed some multiples of 3 and 7 at the right.　　7, 14, 21, 28, 35, 42, 49, 56　　b) 60

(a) Identify the smallest common multiple of both 3 and 7. _____

(b) Therefore, the lowest common denominator for $\frac{2}{3} + \frac{4}{7}$ is _____.

---

64. To find the lowest common denominator for $\frac{5}{6} + \frac{7}{10}$,　　6, 12, 18, 24, 30, 36, 42　　a) 21

we have listed some multiples of 6 and 10 at the right.　　10, 20, 30, 40, 50, 60, 70　　b) 21

(a) Identify the smallest common multiple of both 6 and 10.

(b) Therefore, the lowest common denominator for $\frac{5}{6} + \frac{7}{10}$ is _____.　　　　　　a) 30
　　b) 30

53

65. There is a shortcut that can be used to find the lowest common denominator. Instead of listing the multiples of both denominators, we can simply list the multiples of the larger denominator and look for the smallest one that is a multiple of the smaller denominator. An example is given below.

To find the lowest common denominator for $\frac{5}{6} + \frac{1}{8}$, we have listed some multiples of the larger denominator "8".

8, 16, 24, 32, 40, 48, 56, 64

(a) The smallest multiple of 8 that is also a multiple of 6 is _____.

(b) Therefore, the lowest common denominator for $\frac{5}{6} + \frac{1}{8}$ is _____.

66. To find the lowest common denominator for $\frac{6}{7} + \frac{4}{5}$, we have listed some multiples of the larger denominator "7".

7, 14, 21, 28, 35, 42, 49, 56, 63

(a) The smallest multiple of 7 that is also a multiple of 5 is _____.

(b) Therefore, the lowest common denominator for $\frac{6}{7} + \frac{4}{5}$ is _____.

a) 24
b) 24

67. To find the lowest common denominator for $\frac{7}{9} + \frac{4}{15}$, we have listed some multiples of the larger denominator "15".

15, 30, 45, 60, 75, 90, 105

(a) The smallest multiple of 15 that is also a multiple of 9 is _____.

(b) Therefore, the lowest common denominator for $\frac{7}{9} + \frac{4}{15}$ is _____.

a) 35
b) 35

68. Some multiples of 9 are: 9, 18, 27, 36, 45, 54, 63, 72, 81

By checking the list above, identify the lowest common denominator for each addition below.

(a) $\frac{5}{9} + \frac{1}{4}$ _____   (b) $\frac{5}{6} + \frac{4}{9}$ _____   (c) $\frac{2}{9} + \frac{7}{8}$ _____

a) 45
b) 45

69. Some multiples of 12 are: 12, 24, 36, 48, 60, 72, 84, 96, 108

By checking the list above, identify the lowest common denominator for each addition below.

(a) $\frac{5}{12} + \frac{1}{8}$ _____   (b) $\frac{3}{7} + \frac{1}{12}$ _____   (c) $\frac{11}{12} + \frac{9}{10}$ _____

a) 36   b) 18   c) 72

70. Some multiples of 20 are: 20, 40, 60, 80, 100, 120, 140, 160, 180

By checking the list above, identify the lowest common denominator for each addition below.

(a) $\frac{7}{20} + \frac{1}{6}$ _____   (b) $\frac{1}{12} + \frac{13}{20}$ _____   (c) $\frac{3}{8} + \frac{9}{20}$ _____

a) 24   b) 84   c) 60

a) 60   b) 60   c) 40

54

71. To find a lowest common denominator, we can simply list the multiples of the larger denominator until we find the smallest one that is a multiple of the smaller denominator. An example is given below.

To find the lowest common denominator for $\frac{3}{4} + \frac{1}{10}$, we can list the multiples of 10 until we find the smallest one that is also a multiple of 4.

Is 10 a multiple of 4? No.

Is 20 a multiple of 4? Yes.   Therefore, the lowest common denominator for $\frac{3}{4} + \frac{1}{10}$ is _____.

72. To find the lowest common denominator for $\frac{4}{7} + \frac{2}{3}$, we can list the multiples of 7 until we find the smallest one that is also a multiple of 3.

|20

Is 7 a multiple of 3? No.

Is 14 a multiple of 3? No.

Is 21 a multiple of 3? Yes.   Therefore, the lowest common denominator for $\frac{4}{7} + \frac{2}{3}$ is _____.

73. To find the lowest common denominator for $\frac{3}{10} + \frac{7}{8}$, we can list the multiples of 10 until we find the smallest one that is also a multiple of 8.

|21

Is 10 a multiple of 8? No.

Is 20 a multiple of 8? No.

Is 30 a multiple of 8? No.

Is 40 a multiple of 8? Yes.   Therefore, the lowest common denominator for $\frac{3}{10} + \frac{7}{8}$ is _____.

74. By checking the multiples of the larger denominator, identify the lowest common denominator for each addition below.

|40

(a) $\frac{4}{5} + \frac{2}{3}$ _____   (b) $\frac{7}{12} + \frac{5}{8}$ _____   (c) $\frac{1}{2} + \frac{1}{7}$ _____

75. Identify the lowest common denominator for each addition below.

a) 15 b) 24 c) 14

(a) $\frac{5}{6} + \frac{2}{7}$ _____   (b) $\frac{7}{15} + \frac{1}{10}$ _____   (c) $\frac{3}{4} + \frac{5}{6}$ _____

76. Identify the lowest common denominator for each addition below.

a) 42 b) 30 c) 12

(a) $\frac{7}{10} + \frac{4}{25}$ _____   (b) $\frac{1}{4} + \frac{1}{5}$ _____   (c) $\frac{7}{20} + \frac{5}{6}$ _____

77. If the larger denominator is a multiple of the smaller, the larger denominator itself is the lowest common denominator. For example:

a) 50 b) 20 c) 60

For $\frac{3}{8} + \frac{1}{4}$, the lowest common denominator is 8.   For $\frac{5}{16} + \frac{7}{32}$, the lowest common denominator is _____.

78. Identify the lowest common denominator for each addition below.

|32

(a) $\frac{17}{35} + \frac{1}{7}$ _____   (b) $\frac{7}{8} + \frac{39}{40}$ _____   (c) $\frac{1}{2} + \frac{2}{5}$ _____

79. Identify the lowest common denominator for each addition below.　　　　　　　　　　a) 35　b) 40　c) 10

　　(a) $\frac{3}{4} + \frac{9}{14}$ ____　　(b) $\frac{1}{100} + \frac{1}{10}$ ____　　(c) $\frac{1}{8} + \frac{1}{7}$ ____

80. Identify the lowest common denominator for each addition below.　　　　　　　　　　a) 28　b) 100　c) 56

　　(a) $\frac{5}{6} + \frac{11}{18}$ ____　　(b) $\frac{1}{5} + \frac{3}{9}$ ____　　(c) $\frac{11}{15} + \frac{7}{9}$ ____　　　a) 18　b) 45　c) 45

3-9　ADDING FRACTIONS WHEN THE LARGER DENOMINATOR IS NOT A MULTIPLE OF THE SMALLER

In this section, we will discuss the procedure for adding "unlike" fractions when the larger denominator is not a multiple of the smaller denominator. Lowest common denominators will be used to perform the additions.

---

81. In $\frac{1}{8} + \frac{2}{3}$, 8 is not a multiple of 3. To perform the addition, we use the following steps:

　　(1) Identify the lowest common denominator.　　　　It is 24.

　　(2) Substitute $\frac{3}{24}$ for $\frac{1}{8}$ and $\frac{16}{24}$ for $\frac{2}{3}$ to get　　$\frac{1}{8} + \frac{2}{3} = \frac{3}{24} + \frac{16}{24}$
　　　　an equivalent addition of "like" fractions.

　　(3) Then perform the addition of "like"　　　　$\frac{1}{8} + \frac{2}{3} = \frac{3}{24} + \frac{16}{24} = \frac{19}{24}$
　　　　fractions in the usual way.

　　In $\frac{3}{4} + \frac{1}{6}$, 6 is not a multiple of 4. Let's perform the addition by following the same steps.

　　(a) Identify the lowest common denominator.　　It is ____.

　　(b) Substitute $\frac{9}{12}$ for $\frac{3}{4}$ and $\frac{2}{12}$ for $\frac{1}{6}$ to get　　$\frac{3}{4} + \frac{1}{6} = $ ____ + ____
　　　　an equivalent addition of "like" fractions.

　　(c) Complete the addition.　　$\frac{3}{4} + \frac{1}{6} = $ ____ + ____ = ____

82. (a) The lowest common denominator for the addition below is ____.　　　a) 12

　　(b) Following the steps in the last frame, complete the addition.　　$\frac{1}{2} + \frac{3}{7} = $ ____ + ____ = ____　　b) $\frac{9}{12} + \frac{2}{12}$

　　　　　　　　　　　　　　　　　　　　　　　　　　　　　　　　　　　　　　　c) $\frac{9}{12} + \frac{2}{12} = \frac{11}{12}$

83. (a) The lowest common denominator for the addition below is ____.　　　a) 14　b) $\frac{7}{14} + \frac{6}{14} = \frac{13}{14}$

　　(b) Complete the addition:　　$\frac{5}{8} + \frac{1}{6} = $ ____ + ____ = ____

84. In each case below, identify the lowest common denominator and then complete the addition.　　a) 24　b) $\frac{15}{24} + \frac{4}{24} = \frac{19}{24}$

　　(a) $\frac{4}{7} + \frac{1}{5} = $ ____ + ____ = ____　　(b) $\frac{3}{4} + \frac{1}{10} = $ ____ + ____ = ____

55

85. Complete the additions below. Convert each sum to a mixed number.

   (a) $\frac{2}{3} + \frac{7}{4} =$ ____ + ____ _____

   (b) $\frac{11}{15} + \frac{5}{9} =$ ____ + ____ _____

   a) $\frac{20}{35} + \frac{7}{35} = \frac{27}{35}$ b) $\frac{15}{20} + \frac{2}{20} = \frac{17}{20}$

86. Remember that the larger denominator is the lowest common denominator <u>if it is a multiple of the smaller</u>. Find each sum below.

   (a) $\frac{7}{32} + \frac{7}{8} =$ _____

   (b) $\frac{3}{4} + \frac{8}{9} =$ _____

   a) $\frac{8}{12} + \frac{21}{12} = \frac{29}{12} = 2\frac{5}{12}$ b) $\frac{33}{45} + \frac{25}{45} = \frac{58}{45} = 1\frac{13}{45}$

87. The <u>lowest common denominator</u> for the addition at the right is 18. We used 18 as the common denominator to find the sum.

   $\frac{1}{6} + \frac{7}{9} = \frac{3}{18} + \frac{14}{18} = \frac{17}{18}$

   a) $1\frac{3}{32}$ (from $\frac{35}{32}$)

   b) $1\frac{23}{36}$ (from $\frac{59}{36}$)

   Since 36 and 54 are also multiples of both 6 and 9, we can use either 36 or 54 as the common denominator for the same addition. We have done so below.

   $\frac{1}{6} + \frac{7}{9} = \frac{6}{36} + \frac{28}{36} = \frac{34}{36} = \frac{17}{18}$

   $\frac{1}{6} + \frac{7}{9} = \frac{9}{54} + \frac{42}{54} = \frac{51}{54} = \frac{17}{18}$

   Note: In each of the two additions, we had to reduce the sum to lowest terms.

   Did we obtain the same sum in all three additions above? _____

88. As we saw in the last frame, any number that is a multiple of both denominators can be used as the common denominator for an addition. However, using the <u>lowest common denominator</u> is more efficient for two reasons.

   Yes. It is $\frac{17}{18}$.

   (1) When the lowest common denominator is used, the numbers involved in the addition are smaller.

   (2) When the lowest common denominator is not used, the sum always has to be reduced to lowest terms.

3-10 SUBTRACTING FRACTIONS WHEN THE LARGER DENOMINATOR IS NOT A MULTIPLE OF THE SMALLER

In this section, we will discuss the procedure for subtracting "unlike" fractions when the larger denominator is not a multiple of the smaller denominator. Lowest common denominators will be used to perform the subtractions.

89. In $\frac{1}{2} - \frac{2}{5}$, 5 is not a multiple of 2. To perform the subtraction, we use the same steps that were used in additions of that type.

   (1) Identify the lowest common denominator.   It is 10.

   (2) Substitute $\frac{5}{10}$ for $\frac{1}{2}$ and $\frac{4}{10}$ for $\frac{2}{5}$ to get an equivalent subtraction of "like" fractions.

   $\frac{1}{2} - \frac{2}{5} = \frac{5}{10} - \frac{4}{10}$

   (3) Complete the subtraction in the usual way.

   $\frac{1}{2} - \frac{2}{5} = \frac{5}{10} - \frac{4}{10} = \frac{1}{10}$

   Continued on following page.

89. Continued

In $\frac{7}{8} - \frac{5}{6}$, 8 is not a multiple of 6. Let's use the same steps to perform the subtraction.

(a) Identify the lowest common denominator. It is _____.

(b) Substitute $\frac{21}{24}$ for $\frac{7}{8}$ and $\frac{20}{24}$ for $\frac{5}{6}$ to get an equivalent subtraction of "like" fractions.

$\frac{7}{8} - \frac{5}{6}$ = _____ - _____

(c) Complete the subtraction in the usual way.

$\frac{7}{8} - \frac{5}{6}$ = _____ - _____ = _____

---

90. (a) The lowest common denominator for the subtraction below is _____.

(b) Following the steps in the last frame, complete the subtraction.

$\frac{7}{8} - \frac{2}{3}$ = _____ - _____ = _____

a) It is 24.

b) $\frac{7}{8} - \frac{5}{6} = \frac{21}{24} - \frac{20}{24}$

c) $\frac{7}{8} - \frac{5}{6} = \frac{21}{24} - \frac{20}{24} = \frac{1}{24}$

---

91. (a) The lowest common denominator for the subtraction below is _____.

(b) Complete the subtraction: $\frac{3}{4} - \frac{1}{10}$ = _____ - _____ = _____

a) 24 b) $\frac{21}{24} - \frac{16}{24} = \frac{5}{24}$

---

92. Use the lowest common denominator for each subtraction below.

(a) $\frac{1}{2} - \frac{1}{3}$ = _____ - _____ = _____

(b) $\frac{7}{6} - \frac{2}{9}$ = _____ - _____ = _____

a) 20 b) $\frac{15}{20} - \frac{2}{20} = \frac{13}{20}$

---

93. Use the lowest common denominator for each subtraction below. Write each difference as a mixed number.

(a) $\frac{9}{5} - \frac{1}{6}$ = _____ - _____ = _____

(b) $\frac{33}{20} - \frac{3}{8}$ = _____ - _____ = _____

a) $\frac{3}{6} - \frac{2}{6} = \frac{1}{6}$ b) $\frac{21}{18} - \frac{4}{18} = \frac{17}{18}$

---

94. Remember that the larger denominator is the lowest common denominator <u>if it is a multiple of the smaller</u>. Find each difference below.

(a) $\frac{7}{12} - \frac{1}{3}$ = _____ - _____ = _____

(b) $\frac{15}{8} - \frac{3}{20}$ = _____ - _____ = _____

a) $\frac{54}{30} - \frac{5}{30} = \frac{49}{30} = 1\frac{19}{30}$

b) $\frac{66}{40} - \frac{15}{40} = \frac{51}{40} = 1\frac{11}{40}$

---

95. The lowest common denominator for the subtraction below is 36. We have used it to find the difference.

$\frac{7}{12} - \frac{2}{9} = \frac{21}{36} - \frac{8}{36} = \frac{13}{36}$

a) $\frac{7}{12} - \frac{4}{12} = \frac{3}{12} = \frac{1}{4}$

b) $\frac{75}{40} - \frac{6}{40} = \frac{69}{40} = 1\frac{29}{40}$

Since 72 is also a multiple of both 12 and 9, we can use 72 as the common denominator for the same subtraction. We have done so below.

$\frac{7}{12} - \frac{2}{9} = \frac{42}{72} - \frac{16}{72} = \frac{26}{72} = \frac{13}{36}$

(a) Did we obtain the same difference in each case? _____

(b) Was it easier to use 36 or 72 as the common denominator? _____

57

96. We have used the lowest common denominator to obtain an equivalent subtraction of "like" fractions below. The subtraction cannot be done in arithmetic since 14 is larger than 12.

$$\frac{4}{7} - \frac{2}{3} = \frac{12}{21} - \frac{14}{21}$$

Which of the subtractions at the right cannot be done in arithmetic? _____ (a) $\frac{3}{4} - \frac{2}{3}$ (b) $\frac{5}{8} - \frac{3}{4}$

a) Yes. It is $\frac{13}{36}$.

b) 36, since the numbers were smaller and we did not have to reduce the difference to lowest terms.

(b)

SELF-TEST 6 (Frames 47-96)

Find the lowest common denominator for each of the following.

1. $\frac{17}{20} + \frac{4}{5}$ _____
2. $\frac{5}{8} + \frac{1}{10}$ _____
3. $\frac{11}{12} - \frac{2}{9}$ _____
4. $\frac{6}{7} - \frac{2}{3}$ _____

Do the following additions and subtractions.

5. $\frac{1}{2} + \frac{2}{5} =$
6. $\frac{3}{4} + \frac{5}{6} =$
7. $\frac{7}{15} + \frac{3}{10} =$
8. $\frac{23}{20} + \frac{7}{10} =$

9. $\frac{4}{3} - \frac{1}{2} =$
10. $\frac{7}{4} - \frac{9}{16} =$
11. $\frac{5}{8} - \frac{7}{12} =$
12. $\frac{44}{15} - \frac{5}{6} =$

ANSWERS: 1. 20   3. 36   5. $\frac{9}{10}$ 7. $\frac{23}{30}$ 9. $\frac{5}{6}$ 11. $\frac{1}{24}$

2. 40   4. 21   6. $1\frac{7}{12}$ 8. $1\frac{17}{20}$ 10. $1\frac{3}{16}$ 12. $2\frac{1}{10}$

3-11 ADDING THREE FRACTIONS

In this section, we will discuss the procedure for adding three fractions. Additions involving "like" and "unlike" fractions are included.

97. To add three fractions with "like" denominators, we simply add their numerators. For example:

$$\frac{2}{9} + \frac{1}{9} + \frac{4}{9} = \frac{2+1+4}{9} = \frac{7}{9} \qquad \frac{5}{12} + \frac{1}{12} + \frac{2}{12} = \underline{\qquad}$$

$\frac{2}{3}$ (from $\frac{8}{12}$)

98. In each addition below, the largest denominator is a multiple of the two smaller ones. That is:

In $\frac{3}{4} + \frac{5}{8} + \frac{1}{2}$, 8 is a multiple of both 4 and 2.   In $\frac{2}{3} + \frac{5}{12} + \frac{1}{4}$, 12 is a multiple of both 3 and 4.

When the largest denominator is a multiple of the other two, it is the lowest common denominator. That is:

For $\frac{3}{4} + \frac{5}{8} + \frac{1}{2}$, the lowest common denominator is 8.

For $\frac{2}{3} + \frac{5}{12} + \frac{1}{4}$, the lowest common denominator is _____.

99. For $\frac{1}{4} + \frac{1}{12} + \frac{1}{6}$, the lowest common denominator is 12. The addition is performed by using the following steps. | 12

(1) Substitute $\frac{3}{12}$ for $\frac{1}{4}$ and $\frac{2}{12}$ for $\frac{1}{6}$ to get an equivalent addition of "like" fractions. $\frac{1}{4} + \frac{1}{12} + \frac{1}{6} = \frac{3}{12} + \frac{1}{12} + \frac{2}{12}$

(2) Complete the addition in the usual way. $\frac{1}{4} + \frac{1}{12} + \frac{1}{6} = \frac{3}{12} + \frac{1}{12} + \frac{2}{12} = \frac{6}{12} = \frac{1}{2}$

For the addition at the right, the lowest common denominator is 16. Following the steps above, complete the addition. $\frac{3}{16} + \frac{3}{8} + \frac{1}{4} = \frac{3}{16} + \underline{} + \underline{} = \underline{}$

100. (a) The lowest common denominator for the addition below is _____. | $\frac{3}{16} + \frac{6}{16} + \frac{4}{16} = \frac{13}{16}$

(b) Following the steps in the last frame, complete the addition. Write the sum in lowest terms.

$\frac{1}{4} + \frac{3}{20} + \frac{2}{5} = \underline{} + \underline{} + \underline{} = \underline{}$

101. Perform the addition below. Write the sum in lowest terms. | a) 20   b) $\frac{5}{20} + \frac{3}{20} + \frac{8}{20} = \frac{16}{20} = \frac{4}{5}$

$\frac{7}{15} + \frac{1}{3} + \frac{4}{5} = \underline{} + \underline{} + \underline{} = \underline{}$

102. The larger denominator appears in two fractions in the addition below. Therefore, we only have to make one substitution. That is: | $\frac{7}{15} + \frac{5}{15} + \frac{12}{15} = \frac{24}{15} = 1\frac{3}{5}$

$\frac{3}{8} + \frac{3}{4} + \frac{1}{8} = \frac{3}{8} + \frac{6}{8} + \frac{1}{8} = \frac{10}{8} = 1\frac{1}{4}$

Perform the addition below. Write the sum in lowest terms.

$\frac{7}{16} + \frac{7}{8} + \frac{11}{16} = \underline{} + \underline{} + \underline{} = \underline{}$

| $\frac{7}{16} + \frac{14}{16} + \frac{11}{16} = \frac{32}{16} = 2$

103. In $\frac{3}{5}+\frac{2}{3}+\frac{1}{2}$, 5 is not a multiple of 3 and 2. Therefore, 5 is not the lowest common denominator. To find the lowest common denominator, we check the multiples of 5 until we find the smallest one that is a multiple of both 3 and 2. That is:

    Is 5 a multiple of both 3 and 2? No. Is 20 a multiple of both 3 and 2? No.
    Is 10 a multiple of both 3 and 2? No. Is 25 a multiple of both 3 and 2? No.
    Is 15 a multiple of both 3 and 2? No. Is 30 a multiple of both 3 and 2? Yes.

    Therefore, the lowest common denominator for $\frac{3}{5}+\frac{2}{3}+\frac{1}{2}$ is _____.

104. In $\frac{3}{4}+\frac{5}{6}+\frac{1}{2}$, 6 is not a multiple of both 4 and 2. To find the lowest common denominator, we check the multiples of 6 until we find the smallest one that is a multiple of both 4 and 2. That is:

    | 30 |

    Is 6 a multiple of both 4 and 2? No.
    Is 12 a multiple of both 4 and 2? Yes.

    Therefore, the lowest common denominator for $\frac{3}{4}+\frac{5}{6}+\frac{1}{2}$ is _____.

105. The largest denominator is not a multiple of the other two in the additions below. By checking the multiples of the largest denominator, identify the lowest common denominator in each case.

    | 12 |

    (a) $\frac{3}{2}+\frac{6}{7}+\frac{1}{3}$ _____

    (b) $\frac{5}{12}+\frac{1}{6}+\frac{4}{9}$ _____

106. Check the multiples of the largest denominator to identify the lowest common denominator for each addition below.

    | a) 42   b) 36 |

    (a) $\frac{1}{4}+\frac{4}{5}+\frac{3}{2}$ _____

    (b) $\frac{1}{2}+\frac{5}{7}+\frac{3}{4}$ _____

    | a) 20   b) 28 |

107. In $\frac{3}{4}+\frac{1}{8}+\frac{2}{3}$, 8 is not a multiple of both 4 and 3. The following steps are used to perform the addition.

    (1) Identify the lowest common denominator. It is 24.
    (2) Substitute the fractions whose denominators are 24 and complete the addition. $\frac{3}{4}+\frac{1}{8}+\frac{2}{3} = \frac{18}{24}+\frac{3}{24}+\frac{16}{24} = \frac{37}{24} = 1\frac{13}{24}$

    For the addition at the right, the lowest common denominator is 20. Following the steps above, complete the addition. $\frac{1}{2}+\frac{4}{5}+\frac{1}{4} =$ _____ + _____ + _____ = _____

108. (a) The lowest common denominator for the addition below is _____.

    (b) Following the steps in the last frame, complete the addition.

    $\frac{5}{6}+\frac{3}{8}+\frac{1}{4} =$ _____ + _____ + _____ = _____

    | $\frac{10}{20}+\frac{16}{20}+\frac{5}{20} = \frac{31}{20} = 1\frac{11}{20}$ |

109. (a) The lowest common denominator for the addition below is _____.

    (b) Complete the addition: $\frac{7}{10}+\frac{1}{2}+\frac{3}{4} =$ _____ + _____ + _____ = _____

    | a) 24   b) $\frac{20}{24}+\frac{9}{24}+\frac{6}{24} = \frac{35}{24} = 1\frac{11}{24}$ |

110. Complete:

$\frac{1}{2} + \frac{1}{4} + \frac{1}{3}$ = ___ + ___ + ___ = ___

a) 20 b) $\frac{14}{20} + \frac{10}{20} + \frac{15}{20} = \frac{39}{20} = 1\frac{19}{20}$

$\frac{6}{12} + \frac{3}{12} + \frac{4}{12} = \frac{13}{12} = 1\frac{1}{12}$

3-12 USING THE PRODUCT OF THE DENOMINATORS AS THE COMMON DENOMINATOR

Sometimes it is difficult to identify the lowest common denominator. Since the product of the denominators is always a common denominator, we can use the product as the common denominator in those cases. We will discuss the use of the product of the denominators as the common denominator in this section.

---

111. In any addition or subtraction, the product of the denominators is a multiple of the denominators. That is:

For $\frac{3}{7} + \frac{5}{8}$: the product of the denominators is 56, and 56 is a multiple of both 7 and 8.

For $\frac{5}{9} - \frac{1}{7}$: the product of the denominators is 63, and 63 is a multiple of both 9 and 7.

Since the product of the denominators is a multiple of both denominators, we can use that product as the common denominator.

(a) Use 56 as the common denominator to perform the addition at the right. $\frac{3}{7} + \frac{5}{8}$ = ___ + ___ = ___

(b) Use 63 as the common denominator to perform the subtraction at the right. $\frac{5}{9} - \frac{1}{7}$ = ___ - ___ = ___

112. Sometimes the product of the denominators is the lowest common denominator. Sometimes it is not. For example:

For $\frac{2}{3} + \frac{4}{7}$: the product of the denominators is 21, and 21 is the lowest common denominator.

For $\frac{5}{6} - \frac{3}{4}$: the product of the denominators is 24, and 24 is not the lowest common denominator.

a) $\frac{24}{56} + \frac{35}{56} = \frac{59}{56} = 1\frac{3}{56}$

b) $\frac{35}{63} - \frac{9}{63} = \frac{26}{63}$

In which cases below does the product of the denominators equal the lowest common denominator?

(a) $\frac{3}{5} + \frac{2}{9}$ (b) $\frac{7}{10} - \frac{1}{2}$ (c) $\frac{1}{8} + \frac{5}{6}$ (d) $\frac{5}{9} - \frac{3}{8}$

___

113. Even when the product of the denominators is not the lowest common denominator, we still obtain the same sum if we use that product as the common denominator.

Only (a) and (d)

(a) The product of the denominators for the addition below is 48. Use 48 as the common denominator to find the sum.

$\frac{5}{8} + \frac{1}{6}$ = ___ + ___ = ___

61

113. Continued

    (b) The lowest common denominator for the same addition is 24. Use 24 as the common denominator to find the sum.

$$\frac{5}{8} + \frac{1}{6} = \underline{\hspace{1cm}} + \underline{\hspace{1cm}} = \underline{\hspace{1cm}}$$

    (c) Was the same sum obtained in both cases? _____

---

114. Sometimes it is difficult to identify the lowest common denominator. In such cases, the product of the denominators can be used as the common denominator, whether it is the lowest common denominator or not.

    a) $\frac{30}{48} + \frac{8}{48} = \frac{38}{48} = \frac{19}{24}$

    b) $\frac{15}{24} + \frac{4}{24} = \frac{19}{24}$ c) Yes

Use the product of the denominators as the common denominator to find the sum and difference below.

(a) $\frac{2}{9} + \frac{1}{8} = \underline{\hspace{1cm}} + \underline{\hspace{1cm}} = \underline{\hspace{1cm}}$ (b) $\frac{7}{9} - \frac{3}{10} = \underline{\hspace{1cm}} - \underline{\hspace{1cm}} = \underline{\hspace{1cm}}$

---

115. When the lowest common denominator is smaller than the product of the denominators and easy to find, <u>don't use the product of the denominators as the common denominator</u>. That is:

    a) $\frac{16}{72} + \frac{9}{72} = \frac{25}{72}$ b) $\frac{70}{90} - \frac{27}{90} = \frac{43}{90}$

For $\frac{3}{8} + \frac{1}{4}$, use the lowest common denominator "8" as the common denominator.

For $\frac{9}{10} - \frac{3}{4}$, use the lowest common denominator "20" as the common denominator.

Use the lowest common denominator to find the difference and sum below.

(a) $\frac{7}{10} - \frac{1}{2} = \underline{\hspace{1cm}} - \underline{\hspace{1cm}} = \underline{\hspace{1cm}}$ (b) $\frac{5}{12} + \frac{1}{8} = \underline{\hspace{1cm}} + \underline{\hspace{1cm}} = \underline{\hspace{1cm}}$

---

116. For any addition of three fractions, the product of the denominators is also a multiple of all three denominators. That is:

    a) $\frac{7}{10} - \frac{5}{10} = \frac{2}{10} = \frac{1}{5}$ b) $\frac{10}{24} + \frac{3}{24} = \frac{13}{24}$

For $\frac{1}{2} + \frac{2}{3} + \frac{5}{7}$: the product of the denominators is 42, and 42 is a multiple of 2, 3, and 7.

Since the product of the denominators is a multiple of the three denominators, we can use 42 as the common denominator. Do so below.

$$\frac{1}{2} + \frac{2}{3} + \frac{5}{7} = \underline{\hspace{1cm}} + \underline{\hspace{1cm}} + \underline{\hspace{1cm}} = \underline{\hspace{1cm}}$$

---

117. It is frequently difficult to identify the lowest common denominator for an addition of three fractions. In such cases, the product of the denominators can be used as the common denominator, whether it is the lowest common denominator or not.

    $\frac{21}{42} + \frac{28}{42} + \frac{30}{42} = \frac{79}{42} = 1\frac{37}{42}$

Use the product of the denominators as the common denominator to find the sum below.

$$\frac{1}{2} + \frac{3}{5} + \frac{6}{7} = \underline{\hspace{1cm}} + \underline{\hspace{1cm}} + \underline{\hspace{1cm}} = \underline{\hspace{1cm}}$$

118. When the lowest common denominator is smaller than the product of three denominators and easy to find, don't use the product of the denominators as the common denominator. That is:

    For $\frac{5}{6} + \frac{1}{12} + \frac{2}{3}$, use the lowest common denominator "12" as the common denominator.

    $$\frac{35}{70} + \frac{42}{70} + \frac{60}{70} = \frac{137}{70} = 1\frac{67}{70}$$

    Use the lowest common denominator to find the sum below.

    $\frac{3}{7} + \frac{5}{14} + \frac{1}{2} = $ ____ + ____ + ____ = _____

    $$\frac{6}{14} + \frac{5}{14} + \frac{7}{14} = \frac{18}{14} = 1\frac{2}{7}$$

3-13 COMPARING THE SIZE OF FRACTIONS

In this section, we will discuss a procedure for comparing the size of fractions. In this procedure, "like" or "common" denominators are needed before comparisons can be made.

119. We have already seen that the size of fractions with "like" denominators can be compared by simply comparing the size of their numerators. For example:

    Of $\frac{4}{7}$ and $\frac{6}{7}$, the larger is $\frac{6}{7}$.   Of $\frac{11}{9}$ and $\frac{13}{9}$, the smaller is ____.

120. By comparing the size of the shaded parts in the pair of figures at the left below, you can see that $\frac{2}{3}$ is larger than $\frac{4}{9}$. Encircle the larger fraction in each of the other two pairs.

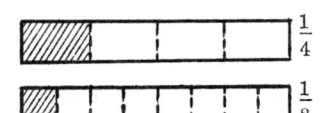

    $\frac{11}{9}$

121. In the last frame, we saw the three facts at the right. By examining them, you can see the following:

    If two fractions have "unlike" denominators, we cannot compare their size by simply comparing the size of their numerators.

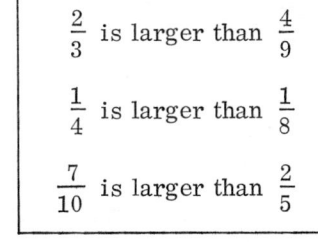

    $\frac{2}{3}$ is larger than $\frac{4}{9}$

    $\frac{1}{4}$ is larger than $\frac{1}{8}$

    $\frac{7}{10}$ is larger than $\frac{2}{5}$

    $\frac{1}{4}$

    $\frac{7}{10}$

    (a) Can we decide that $\frac{7}{12}$ is larger than $\frac{5}{8}$ simply because 7 is larger than 5? _____

    (b) Can we decide that $\frac{5}{6}$ is smaller than $\frac{7}{10}$ simply because 5 is smaller than 7? _____

    a) No   b) No

122. When fractions have "unlike" denominators, we must get common denominators first before comparing their size by simply comparing the size of their numerators. For example:

To decide whether $\frac{5}{8}$ or $\frac{9}{16}$ is larger, we got common denominators at the left below.

$\frac{5}{8}$ $\frac{9}{16}$
↓   ↓   Since $\frac{10}{16}$ is larger than $\frac{9}{16}$, $\frac{5}{8}$ is larger than $\frac{9}{16}$.
$\frac{10}{16}$ $\frac{9}{16}$

To decide whether $\frac{2}{3}$ or $\frac{3}{4}$ is larger, we got common denominators at the left below.

$\frac{2}{3}$ $\frac{3}{4}$
↓   ↓   Since $\frac{9}{12}$ is larger than $\frac{8}{12}$, $\frac{3}{4}$ is larger than $\frac{2}{3}$.
$\frac{8}{12}$ $\frac{9}{12}$

For each pair of fractions at the right, get common denominators and then underline the <u>larger</u> fraction in each pair.   (a) $\frac{7}{16}$ or $\frac{3}{8}$ (b) $\frac{7}{10}$ or $\frac{5}{6}$

123. For each pair of fractions at the right, get common denominators and then underline the <u>smaller</u> fraction in each pair.   (a) $\frac{3}{4}$ or $\frac{4}{5}$ (b) $\frac{5}{8}$ or $\frac{7}{12}$

a) $\frac{7}{16}$ b) $\frac{5}{6}$

124. The three fractions $\frac{5}{7}$, $\frac{1}{7}$, and $\frac{3}{7}$ have "like" denominators.

Of the three: (a) the <u>largest</u> is _____   (b) the <u>smallest</u> is _____

a) $\frac{3}{4}$ b) $\frac{7}{12}$

125. The three fractions at the right have "unlike" denominators. To identify the <u>largest</u> of the three, we got common denominators first.

The <u>largest</u> is $\frac{7}{8}$.

$\frac{3}{4}$ $\frac{7}{8}$ $\frac{11}{16}$
↓   ↓   ↓
$\frac{12}{16}$ $\frac{14}{16}$ $\frac{11}{16}$

After getting common denominators, underline the <u>largest</u> of the three fractions in each case at the right.   (a) $\frac{1}{2}$, $\frac{3}{4}$, $\frac{5}{8}$ (b) $\frac{4}{5}$, $\frac{7}{10}$, $\frac{13}{20}$

a) $\frac{5}{7}$ b) $\frac{1}{7}$

126. The three fractions at the right have "unlike" denominators. To identify the <u>smallest</u> of the three, we got common denominators first.

The <u>smallest</u> is $\frac{1}{2}$.

$\frac{1}{2}$ $\frac{3}{5}$ $\frac{2}{3}$
↓   ↓   ↓
$\frac{15}{30}$ $\frac{18}{30}$ $\frac{20}{30}$

a) $\frac{3}{4}$ b) $\frac{4}{5}$

Get common denominators and then underline the <u>smallest</u> of the three fractions in each case at the right.   (a) $\frac{1}{2}$, $\frac{2}{3}$, $\frac{3}{4}$ (b) $\frac{3}{4}$, $\frac{7}{8}$, $\frac{7}{10}$

a) $\frac{1}{2}$ b) $\frac{7}{10}$

3-14   APPLIED PROBLEMS

This section contains some verbal or applied problems. All of them can be solved by either adding, subtracting, or comparing the size of fractions.

The first four problems can be solved by an addition of fractions.

127. A girl grew $\frac{3}{4}$ of an inch in the first six months of a year. She grew another $\frac{7}{8}$ of an inch in the last six months of that year. Find her total increase in height for the whole year.

$1\frac{5}{8}$ inches

128. A boy rode $\frac{7}{10}$ of a mile to a store on his bicycle. He then rode $\frac{9}{10}$ of a mile to his friend's house. Find the total distance traveled.

$1\frac{3}{5}$ miles

129. A recipe requires $\frac{3}{4}$ of a cup of milk and $\frac{1}{3}$ of a cup of water. Find the total amount of liquid required.

$1\frac{1}{12}$ cups

130. A woman bought $\frac{3}{4}$ of a pound of caramels, $\frac{1}{2}$ of a pound of mints, and $\frac{3}{4}$ of a pound of fudge. Find the total weight of the candy bought.

2 pounds

The next four problems can be solved by a subtraction of fractions.

131. A man bought 2 lots. The first was $\frac{2}{3}$ of an acre. The second was $\frac{7}{8}$ of an acre. How much larger was the second lot?

$\frac{5}{24}$ acre

132. Bill walked 3 miles in $\frac{5}{6}$ of an hour. Joe walked the same 3 miles in $\frac{2}{3}$ of an hour. How much longer did it take Bill?

$\frac{1}{6}$ hour

133. During the same summer, Mary grew $\frac{7}{8}$ of an inch and Joan grew $\frac{1}{2}$ of an inch. How much more did Mary grow?

$\frac{3}{8}$ inch

134. One recipe requires $\frac{2}{3}$ of a cup of milk and a second recipe requires only $\frac{1}{2}$ of a cup of milk. How much more milk is required by the first recipe?

$\frac{1}{6}$ cup

The next two problems can be solved by comparing the size of the fractions.

135. A man bought two lots. Lot #1 was $\frac{7}{8}$ of an acre. Lot #2 was $\frac{9}{10}$ of an acre. Which lot was larger?

Lot #2

136. Which wrench is smaller, a $\frac{3}{4}$-inch wrench or a $\frac{13}{16}$-inch wrench?

$\frac{3}{4}$-inch

SELF-TEST 7 (Frames 97-136)

Do the following additions.

1. $\frac{2}{3} + \frac{5}{6} + \frac{3}{2} =$

2. $\frac{4}{5} + \frac{1}{2} + \frac{3}{4} =$

3. Which fraction below is largest?

   $\frac{3}{4}$ $\frac{5}{6}$ $\frac{9}{10}$

4. Which fraction below is smallest?

   $\frac{5}{12}$ $\frac{3}{8}$ $\frac{1}{2}$

5. A board whose thickness is $\frac{5}{8}$ of an inch is placed on a board whose thickness is $\frac{13}{16}$ of an inch. Find the total thickness of the two boards.

6. Two rolls of masking tape have widths of $\frac{7}{8}$ of an inch and $\frac{1}{2}$ of an inch. Find the difference in their widths.

ANSWERS:

1. 3
2. $2\frac{1}{20}$
3. $\frac{9}{10}$
4. $\frac{3}{8}$
5. $1\frac{7}{16}$ inches
6. $\frac{3}{8}$ of an inch

Unit 4 MULTIPLICATION AND DIVISION OF FRACTIONS

In this unit, we will discuss multiplications and divisions involving fractions. A special section is devoted to contrasting multiplications and divisions with additions and subtractions of fractions. A special section is also devoted to applied problems involving all four operations with fractions.

4-1 THE PROCEDURE FOR MULTIPLYING TWO FRACTIONS

In this section, we will simply show the procedure for multiplying two fractions. The procedure will be justified in the next section.

1. A multiplication of two fractions is shown at the right. As you can see from the example, the following steps are used to multiply two fractions.

    $$\frac{2}{3} \times \frac{5}{7} = \frac{2 \times 5}{3 \times 7} = \frac{10}{21}$$

    (1) Multiply the two numerators.
    (2) Multiply the two denominators.

    Using the two steps above, complete each multiplication below.

    (a) $\frac{1}{3} \times \frac{1}{4} = \frac{1 \times 1}{3 \times 4} =$ _____   (b) $\frac{3}{5} \times \frac{7}{8} =$ _____   (c) $\frac{2}{3} \times \frac{7}{5} =$ _____

2. When two fractions are multiplied, the answer is called the "product". If the product is a proper fraction, it is always reduced to lowest terms. For example:

    $\frac{1}{4} \times \frac{4}{5} = \frac{1 \times 4}{4 \times 5} = \frac{4}{20} = \frac{1}{5}$ $\frac{3}{5} \times \frac{5}{4} = \frac{3 \times 5}{5 \times 4} =$ _____ $=$ _____

    a) $\frac{1}{12}$ b) $\frac{21}{40}$ c) $\frac{14}{15}$

3. Products can also be improper fractions. For example:

    $\frac{2}{3} \times \frac{11}{7} = \frac{2 \times 11}{3 \times 7} = \frac{22}{21}$ $\frac{3}{2} \times \frac{7}{5} = \frac{3 \times 7}{2 \times 5} =$ _____

    $= \frac{15}{20} = \frac{3}{4}$

4. Some improper-fraction products reduce to whole numbers. For example:

    $\frac{8}{5} \times \frac{5}{8} = \frac{8 \times 5}{5 \times 8} = \frac{40}{40} = 1$ $\frac{3}{2} \times \frac{10}{3} = \frac{3 \times 10}{2 \times 3} =$ _____

    $\frac{21}{10}$

    5 (from $\frac{30}{6}$)

5. If an improper-fraction product does not reduce to a whole number, it is converted to a mixed number. For example:

$$\frac{5}{6} \times \frac{5}{3} = \frac{5 \times 5}{6 \times 3} = \frac{25}{18} = 1\frac{7}{18} \qquad\qquad \frac{7}{4} \times \frac{7}{6} = \frac{7 \times 7}{4 \times 6} = \frac{49}{24} = \underline{\qquad}$$

6. When an improper-fraction product is converted to a mixed number, the fraction part is always reduced to lowest terms. For example:

$$2\frac{1}{24}$$

$$\frac{3}{2} \times \frac{5}{3} = \frac{3 \times 5}{2 \times 3} = \frac{15}{6} = 2\frac{3}{6} = 2\frac{1}{2} \qquad\qquad \frac{11}{5} \times \frac{5}{8} = \frac{11 \times 5}{5 \times 8} = \frac{55}{40} = \underline{\qquad}$$

7. Report each product below in lowest terms.

$$1\frac{3}{8} \text{ (from } 1\frac{15}{40}\text{)}$$

(a) $\frac{1}{4} \times \frac{1}{5} = \underline{\qquad}$ (b) $\frac{10}{3} \times \frac{3}{10} = \underline{\qquad}$ (c) $\frac{5}{3} \times \frac{9}{7} = \underline{\qquad}$

8. Report each product below in lowest terms.

a) $\frac{1}{20}$ b) 1 c) $2\frac{1}{7}$

(a) $\frac{8}{3} \times \frac{1}{10} = \underline{\qquad}$ (b) $\frac{5}{6} \times \frac{7}{2} = \underline{\qquad}$ (c) $\frac{12}{5} \times \frac{5}{2} = \underline{\qquad}$

9. In any multiplication of two fractions, the two fractions multiplied are called the "factors". That is:

a) $\frac{4}{15}$ b) $2\frac{11}{12}$ c) 6

In $\frac{1}{2} \times \frac{3}{5} = \frac{3}{10}$, $\frac{1}{2}$ and $\frac{3}{5}$ are called the "factors".

In $\frac{5}{4} \times \frac{9}{7} = \frac{45}{28}$, \underline{\qquad} and \underline{\qquad} are called the "factors".

10. The "order" principle also applies to multiplications of two fractions. That is: In any multiplication of two fractions, we can interchange the factors without changing the product. For example:

$\frac{5}{4}$ and $\frac{9}{7}$

$\frac{2}{3} \times \frac{5}{7}$ and $\frac{5}{7} \times \frac{2}{3}$ are equal, since the product for each is $\frac{10}{21}$.

$\frac{1}{2} \times \frac{7}{8}$ and $\frac{7}{8} \times \frac{1}{2}$ are equal, since the product for each is \underline{\qquad}.

11. When saying a multiplication of fractions in words, either the word "times" or the word "of" can be used for the multiplication symbol. For example:

$\frac{7}{16}$

$\frac{1}{2} \times \frac{1}{4}$ can be said: "$\frac{1}{2}$ times $\frac{1}{4}$" or "$\frac{1}{2}$ of $\frac{1}{4}$"

Write each product below in lowest terms.

(a) $\frac{1}{3}$ of $\frac{1}{2} = \underline{\qquad}$ (b) $\frac{3}{4}$ times $\frac{4}{3} = \underline{\qquad}$ (c) $\frac{1}{5}$ of $\frac{20}{3} = \underline{\qquad}$

a) $\frac{1}{6}$ b) 1 c) $1\frac{1}{3}$

4-2 JUSTIFYING THE PROCEDURE FOR MULTIPLYING TWO FRACTIONS

In this section, we will briefly justify the procedure for multiplying two fractions by means of some diagrams.

12. The multiplication $\frac{1}{2} \times \frac{2}{3} = \frac{2}{6} = \frac{1}{3}$ is shown in the figures below.

    The top figure represents the fraction $\frac{2}{3}$.

    The bottom figure represents $\frac{1}{2}$ of $\frac{2}{3}$ or $\frac{1}{2} \times \frac{2}{3}$.

    The darker part of the bottom figure represents the product of the multiplication. Compared to the total bottom figure, it represents $\frac{1}{3}$. Is this the same product we got using the ordinary multiplication procedure above? _____

    Yes

13. The multiplication $\frac{2}{3} \times \frac{6}{7} = \frac{12}{21} = \frac{4}{7}$ is shown in the figures below.

    The top figure represents the fraction $\frac{6}{7}$.

    The bottom figure represents $\frac{2}{3}$ of $\frac{6}{7}$ or $\frac{2}{3} \times \frac{6}{7}$.

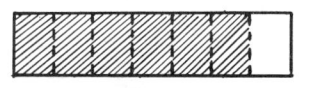

    The darker part of the bottom figure represents the product of the multiplication. Compared to the total bottom figure, it represents $\frac{4}{7}$. Is this the same product we got using the ordinary multiplication procedure above? _____

    Yes

14. The multiplication $\frac{3}{4} \times \frac{4}{5} = \frac{12}{20} = \frac{3}{5}$ is shown in the figures below.

    The top figure represents the fraction $\frac{4}{5}$.

    The bottom figure represents $\frac{3}{4}$ of $\frac{4}{5}$ or $\frac{3}{4} \times \frac{4}{5}$.

    The darker part of the bottom figure represents the product of the multiplication. Compared to the total bottom figure, it represents $\frac{3}{5}$. Is this the same product we got using the ordinary multiplication procedure above? _____

    Yes

4-3 MULTIPLICATIONS INVOLVING A FRACTION AND A WHOLE NUMBER

In this section, we will discuss the procedure for multiplications involving a fraction and a whole number. We will show that any multiplication of that type can be converted to a multiplication of two fractions.

15. Any multiplication involving a fraction and a whole number can be converted to a multiplication involving two fractions. To do so, we simply substitute a fraction for the whole number. For example:

$$7 \times \frac{3}{8} = \frac{7}{1} \times \frac{3}{8} \qquad \frac{5}{6} \times 3 = \frac{5}{6} \times \frac{3}{1}$$

Note: We substituted $\frac{7}{1}$ for 7 and $\frac{3}{1}$ for 3.

Convert each multiplication at the right into a multiplication of two fractions. (a) $8 \times \frac{4}{5} = $ ___ x ___ (b) $\frac{1}{6} \times 5 = $ ___ x ___

---

16. Having converted a multiplication of a fraction and a whole number into a multiplication of two fractions, we can perform the multiplication in the usual way. For example:

a) $\frac{8}{1} \times \frac{4}{5}$ b) $\frac{1}{6} \times \frac{5}{1}$

$$3 \times \frac{5}{8} = \frac{3}{1} \times \frac{5}{8} = \frac{3 \times 5}{1 \times 8} = \frac{15}{8} = 1\frac{7}{8} \qquad \frac{1}{9} \times 7 = \frac{1}{9} \times \frac{7}{1} = \frac{1 \times 7}{9 \times 1} = \frac{7}{9}$$

Convert each of the following to a multiplication of two fractions and then find the product.

(a) $5 \times \frac{2}{7} = $ ___ x ___ = ___   (b) $\frac{3}{4} \times 9 = $ ___ x ___ = ___

---

17. Convert each of the following to a multiplication of two fractions and then find the product.

a) $\frac{5}{1} \times \frac{2}{7} = \frac{10}{7} = 1\frac{3}{7}$ b) $\frac{3}{4} \times \frac{9}{1} = \frac{27}{4} = 6\frac{3}{4}$

(a) $7 \times \frac{6}{7} = $ ___ x ___ = ___   (b) $\frac{1}{8} \times 12 = $ ___ x ___ = ___

---

18. We have performed each multiplication below by converting the multiplication to a multiplication of two fractions.

a) $\frac{7}{1} \times \frac{6}{7} = \frac{42}{7} = 6$ b) $\frac{1}{8} \times \frac{12}{1} = \frac{12}{8} = 1\frac{1}{2}$

$$3 \times \frac{2}{7} = \frac{3}{1} \times \frac{2}{7} = \frac{6}{7} \qquad \frac{2}{13} \times 5 = \frac{2}{13} \times \frac{5}{1} = \frac{10}{13}$$

There is a shortcut that can be used for the same multiplications. The shortcut is shown below.

$$3 \times \frac{2}{7} = \frac{3 \times 2}{7} = \frac{6}{7} \qquad \frac{2}{13} \times 5 = \frac{2 \times 5}{13} = \frac{10}{13}$$

Notice the two steps used in the shortcut:

(1) We multiplied the <u>whole number</u> and <u>the original numerator</u> to get the <u>numerator of the product</u>.

(2) We used the <u>original denominator</u> as the <u>denominator of the product</u>.

Use the shortcut to find each product at the right. (a) $5 \times \frac{1}{8} = \frac{5 \times 1}{8} = $ ___   (b) $\frac{3}{10} \times 3 = \frac{3 \times 3}{10} = $ ___

---

19. Use the shortcut for each multiplication below. Write each product in lowest terms.

a) $\frac{5}{8}$ b) $\frac{9}{10}$

(a) $2 \times \frac{3}{8} = $ ___   (b) $\frac{5}{9} \times 6 = $ ___   (c) $10 \times \frac{1}{5} = $ ___

a) $\frac{3}{4}$ b) $3\frac{1}{3}$ c) 2

20. Write each product below in lowest terms.

   (a) $\frac{3}{5} \times 5 =$ _____   (b) $10 \times \frac{2}{7} =$ _____   (c) $\frac{1}{4} \times 16 =$ _____

21. The "order" principle also applies to multiplications involving a fraction and a whole number. That is:

   $3 \times \frac{1}{7}$ and $\frac{1}{7} \times 3$ are equal, since the product for each is $\frac{3}{7}$.

   $\frac{2}{11} \times 5$ and $5 \times \frac{2}{11}$ are equal, since the product for each is _____.

   a) 3   b) $2\frac{6}{7}$ c) 4

22. When saying a multiplication of a fraction by a whole number in words, we only use the word "times" for the multiplication symbol. For example:

   $5 \times \frac{2}{3}$ is only said: "5 times $\frac{2}{3}$"

   When saying a multiplication of a whole number by a fraction in words, we can use either "times" or "of" for the multiplication symbol. For example:

   $\frac{1}{8} \times 7$ can be said: "$\frac{1}{8}$ times 7" or "$\frac{1}{8}$ of 7".

   Write each product below in lowest terms.

   (a) $\frac{5}{6}$ of $10 =$ _____   (b) 20 times $\frac{1}{2} =$ _____   (c) $\frac{4}{3}$ times $9 =$ _____

   $\frac{10}{11}$

   a) $8\frac{1}{3}$ b) 10   c) 12

4-4 MULTIPLICATIONS INVOLVING A FRACTION AND EITHER "1" OR "0"

In this section, we will discuss multiplications involving a fraction and either "1" or "0".

23. We have used the shortcut to multiply each fraction below by "1".

   $1 \times \frac{5}{8} = \frac{1 \times 5}{8} = \frac{5}{8}$ $\frac{4}{7} \times 1 = \frac{4 \times 1}{7} = \frac{4}{7}$

   From the examples, you can see this fact: <u>When any fraction is multiplied by "1", the product is identical to the original fraction.</u> Using that fact, write each product below.

   (a) $\frac{6}{7} \times 1 =$ _____   (b) $1 \times \frac{9}{10} =$ _____   (c) $1 \times \frac{1}{4} =$ _____   (d) $\frac{1}{8} \times 1 =$ _____

24. When an improper fraction is multiplied by "1", the product is the original improper fraction. We convert this product to a mixed number. That is:

   a) $\frac{6}{7}$ b) $\frac{9}{10}$ c) $\frac{1}{4}$ d) $\frac{1}{8}$

   $1 \times \frac{4}{3} = \frac{4}{3} = 1\frac{1}{3}$ (a) $\frac{21}{8} \times 1 = \frac{21}{8} =$ _____   (b) $1 \times \frac{11}{2} =$ _____

   a) $2\frac{5}{8}$ b) $5\frac{1}{2}$

25. We have used the shortcut to multiply each fraction below by "0".

$$0 \times \frac{3}{8} = \frac{0 \times 3}{8} = \frac{0}{8} = 0 \qquad \frac{7}{9} \times 0 = \frac{7 \times 0}{9} = \frac{0}{9} = 0$$

From the examples, you can see this fact: <u>When any fraction is multiplied by "0", the product is "0"</u>.
Using this fact, write each product below.

(a) $\frac{1}{10} \times 0 = $ _____   (b) $0 \times \frac{9}{2} = $ _____   (c) $\frac{4}{5} \times 0 = $ _____

26. Complete: (a) $0 \times \frac{2}{5} = $ _____   (b) $1 \times \frac{2}{5} = $ _____   (c) $1 \times \frac{5}{2} = $ _____ | a) 0   b) 0   c) 0

27. Complete: (a) $\frac{11}{4} \times 1 = $ _____   (b) $\frac{11}{4} \times 0 = $ _____   (c) $\frac{4}{11} \times 0 = $ _____ | a) 0   b) $\frac{2}{5}$ c) $2\frac{1}{2}$

| a) $2\frac{3}{4}$ b) 0   c) 0

4-5 USING PARENTHESES AS THE MULTIPLICATION SYMBOL

Up to this point, we have used the letter "x" as the multiplication symbol in multiplications involving fractions. In this section, we will show that parentheses can also be used as the multiplication symbol.

28. Parentheses can also be used as the multiplication symbol when fractions are involved. For example:

Both $\left(\frac{3}{8}\right)\left(\frac{1}{2}\right)$ and $\frac{3}{8}\left(\frac{1}{2}\right)$ mean "$\frac{3}{8}$ times $\frac{1}{2}$"   Both $(5)\left(\frac{1}{4}\right)$ and $5\left(\frac{1}{4}\right)$ mean "5 times $\frac{1}{4}$"

Write each product below in lowest terms.

(a) $\left(\frac{1}{2}\right)\left(\frac{10}{11}\right) = $ _____   (b) $2\left(\frac{1}{2}\right) = $ _____   (c) $\frac{7}{8}\left(\frac{8}{7}\right) = $ _____

29. In $2\left(\frac{3}{5}\right)$, the <u>parentheses</u> indicate a multiplication. Don't confuse the multiplication $2\left(\frac{3}{5}\right)$ with the mixed number $2\frac{3}{5}$. | a) $\frac{5}{11}$ b) 1   c) 1

Though $2\left(\frac{3}{5}\right)$ means "2 times $\frac{3}{5}$", $2\frac{3}{5}$ means "2 <u>plus</u> $\frac{3}{5}$".

Which of the following are mixed numbers? _____   (a) $3\left(\frac{7}{11}\right)$ (b) $5\left(\frac{4}{9}\right)$ (c) $7\frac{1}{3}$ (d) $2\left(\frac{1}{8}\right)$

30. Which of the following are multiplications? _____ | Only (c)

(a) $2\frac{1}{2}$ (b) $5\left(\frac{3}{4}\right)$ (c) $4\frac{7}{8}$ (d) $10\frac{5}{16}$

31. (a) Does $3\frac{4}{7}$ mean "3 times $\frac{4}{7}$"? _____   (c) Does $7\left(\frac{2}{3}\right)$ mean "$7 \times \frac{2}{3}$"? _____ | Only (b)

(b) Does $5\left(\frac{3}{8}\right)$ mean "add 5 and $\frac{3}{8}$"? _____   (d) Does $1\frac{3}{4}$ mean "$1 + \frac{3}{4}$"? _____

| a) No   b) No   c) Yes   d) Yes

4-6   THE "CANCELLING" PROCESS FOR MULTIPLICATIONS INVOLVING FRACTIONS

There is a shortcut called the "cancelling" process that can be used to simplify some multiplications involving fractions. We will discuss the "cancelling" process in this section.

32. In the multiplication at the right, the product had to be reduced to lowest terms.

$$\frac{4}{7} \times \frac{3}{8} = \frac{12}{56} = \frac{3}{14}$$

The need to reduce the product to lowest terms can be avoided by a process called "cancelling". The "cancelling" process for the same multiplication is shown at the right and described below:

$$\frac{\overset{1}{\cancel{4}}}{7} \times \frac{3}{\underset{2}{\cancel{8}}} = \frac{3}{14}$$

(1) We divided the numerator "4" and the denominator "8" by 4.

(2) To get the numerator of the product, we multiplied "1" and "3".

(3) To get the denominator of the product, we multiplied "7" and "2".

Let's use the "cancelling" process for each multiplication below.

(a) Divide both the "6" and "8" by 2 before multiplying. $\frac{6}{7} \times \frac{5}{8} =$ _____

(b) Divide both the "9" and "15" by 3 before multiplying. $\frac{7}{15} \times \frac{9}{8} =$ _____

33. The product at the right had to be reduced to lowest terms.

$$\left(\frac{15}{8}\right)\left(\frac{4}{9}\right) = \frac{60}{72} = \frac{5}{6}$$

That reduction can be avoided by using the "cancelling" process. We have done so at the right. The steps are described below.

$$\left(\frac{\overset{5}{\cancel{15}}}{\underset{2}{\cancel{8}}}\right)\left(\frac{\overset{1}{\cancel{4}}}{\underset{3}{\cancel{9}}}\right) = \frac{5}{6}$$

a) $\frac{15}{28}$, from $\frac{\overset{3}{\cancel{6}}}{7} \times \frac{5}{\underset{}{\cancel{8}}} = \frac{15}{28}$

b) $\frac{21}{40}$, from $\frac{7}{\underset{5}{\cancel{15}}} \times \frac{\overset{3}{\cancel{9}}}{8} = \frac{21}{40}$

(1) We divided both 15 and 9 by 3.

(2) We divided both 4 and 8 by 4.

(3) We found the numerator and denominator of the product by multiplying "5 times 1" and "2 times 3".

Let's use the "cancelling" process for each multiplication below.

(a) Divide both the "4" and "10" by 2 and the "7" and "21" by 7 before multiplying. $\frac{4}{21} \times \frac{7}{10} =$ _____

(b) Divide both the "10" and "25" by 5 and the "18" and "27" by 9 before multiplying. $\left(\frac{10}{27}\right)\left(\frac{18}{25}\right) =$ _____

34. We used the "cancelling" process to simplify the multiplication at the right. Notice that the numerator of the product was obtained by multiplying "1" and "1".

$$\frac{\overset{1}{\cancel{3}}}{\underset{2}{\cancel{10}}} \times \frac{\overset{1}{\cancel{5}}}{\underset{3}{\cancel{9}}} = \frac{1}{6}$$

a) $\frac{2}{15}$, from $\frac{\overset{2}{\cancel{4}}}{\underset{3}{\cancel{21}}} \times \frac{\overset{1}{\cancel{7}}}{\underset{5}{\cancel{10}}} = \frac{2}{15}$

b) $\frac{4}{15}$, from $\left(\frac{\overset{2}{\cancel{10}}}{\underset{3}{\cancel{27}}}\right)\left(\frac{\overset{2}{\cancel{18}}}{\underset{5}{\cancel{25}}}\right) = \frac{4}{15}$

Use "cancelling" to simplify each multiplication below before multiplying.

(a) $\frac{2}{21} \times \frac{7}{10} =$ _____   (b) $\left(\frac{11}{18}\right)\left(\frac{9}{44}\right) =$ _____

74

35. We used the "cancelling" process to simplify the multiplication at the right. Notice these points:

$$\frac{\cancel{6}^2}{\cancel{7}_1} \times \frac{\cancel{7}^1}{\cancel{3}_1} = \frac{2}{1} = 2$$

(1) The denominator of the product was obtained by multiplying "1" and "1".

(2) Since the denominator of the product is "1", the product was written as a whole number.

Use "cancelling" to simplify each of these:

(a) $\dfrac{12}{5} \times \dfrac{25}{4} = $ _____

(b) $\left(\dfrac{17}{13}\right)\left(\dfrac{13}{17}\right) = $ _____

a) $\dfrac{1}{15}$, from $\dfrac{\cancel{2}^1}{\cancel{21}_3} \times \dfrac{\cancel{7}^1}{\cancel{10}_5} = \dfrac{1}{15}$

b) $\dfrac{1}{8}$, from $\left(\dfrac{\cancel{11}^1}{\cancel{16}_2}\right)\left(\dfrac{\cancel{8}^1}{\cancel{11}_4}\right) = \dfrac{1}{8}$

36. The product at the right had to be reduced to lowest terms.

$$4 \times \frac{7}{8} = \frac{28}{8} = 3\frac{4}{8} = 3\frac{1}{2}$$

The reduction to lowest terms can be avoided by cancelling. We have done so at the right. Notice that we divided both the "4" and "8" by 4 before multiplying.

$$\cancel{4}^1 \times \frac{7}{\cancel{8}_2} = \frac{7}{2} = 3\frac{1}{2}$$

Use cancelling to simplify each multiplication at the right.

(a) $3 \times \dfrac{7}{9} = $ _____

(b) $10\left(\dfrac{5}{8}\right) = $ _____

a) 15, from $\dfrac{\cancel{12}^3}{\cancel{5}_1} \times \dfrac{\cancel{25}^5}{\cancel{4}_1} = \dfrac{15}{1}$

b) 1, from $\left(\dfrac{\cancel{17}^1}{\cancel{13}_1}\right)\left(\dfrac{\cancel{13}^1}{\cancel{17}_1}\right) = \dfrac{1}{1}$

37. We have used cancelling to simplify each multiplication below. Notice in each case that the product equals a whole number since its denominator is "1".

$$\frac{6}{\cancel{7}_1} \times \cancel{7}^1 = \frac{6}{1} = 6$$

$$\cancel{14}^2\left(\frac{6}{\cancel{7}_1}\right) = \frac{12}{1} = 12$$

Use cancelling to simplify each multiplication at the right.

(a) $\dfrac{5}{8} \times 24 = $ _____

(b) $17\left(\dfrac{13}{17}\right) = $ _____

a) $2\dfrac{1}{3}$, from $\dfrac{\cancel{3}^1}{1} \times \dfrac{7}{\cancel{9}_3} = \dfrac{7}{3}$

b) $6\dfrac{1}{4}$, from $\cancel{10}^5\left(\dfrac{5}{\cancel{8}_4}\right) = \dfrac{25}{4}$

38. Ordinarily the fractions in a multiplication are in lowest terms. However, if one or both are not in lowest terms, we can use cancelling to reduce to lowest terms before multiplying. An example is given at the left below. Complete the other two problems.

a) 15, from $\dfrac{5}{\cancel{8}_1} \times \cancel{24}^3 = \dfrac{15}{1}$

b) 13, from $\cancel{17}^1\left(\dfrac{13}{\cancel{17}_1}\right) = \dfrac{13}{1}$

$$\left(\frac{\cancel{5}^1}{\cancel{10}_2}\right)\left(\frac{7}{9}\right) = \frac{7}{18}$$

(a) $\dfrac{2}{6} \times \dfrac{4}{16} = $ _____

(b) $7\left(\dfrac{9}{12}\right) = $ _____

a) $\dfrac{1}{12}$, from $\dfrac{\cancel{2}^1}{\cancel{6}_3} \times \dfrac{\cancel{4}^1}{\cancel{16}_4} = \dfrac{1}{12}$

b) $5\dfrac{1}{4}$, from $7\left(\dfrac{\cancel{9}^3}{\cancel{12}_4}\right) = \dfrac{21}{4}$

39. Cancelling is not possible with all multiplications involving fractions. For example, cancelling is not possible with any of the multiplications below.

$$\frac{7}{8} \times \frac{5}{9} \qquad 36\left(\frac{3}{5}\right) \qquad \frac{5}{7} \times 13$$

However, since cancelling is a process that obviously simplifies multiplications involving fractions, use it whenever possible.

4-7 APPLIED PROBLEMS

This section contains some verbal or applied problems. All of them can be solved by a multiplication that contains a fraction or fractions.

40. When a car's gas tank is full, it contains 26 gallons. How many gallons does it contain when it is $\frac{1}{4}$ full? (Think: $\frac{1}{4}$ of 26 or $\frac{1}{4} \times 26$)

$6\frac{1}{2}$ gallons

41. If a dozen eggs cost 84 cents, find the cost of $\frac{3}{4}$ of a dozen. (Think: $\frac{3}{4}$ of 84 or $\frac{3}{4} \times 84$)

63 cents

42. A recipe for one apple pie requires $\frac{2}{3}$ of a cup of milk. How many cups of milk will be needed to bake 5 apple pies? (Think: 5 times $\frac{2}{3}$ or $5 \times \frac{2}{3}$)

$3\frac{1}{3}$ cups

43. When making a dress, a woman started with $\frac{2}{3}$ of a yard of ribbon and used $\frac{3}{8}$ of it. How much ribbon did she use? (Think: $\frac{3}{8}$ of $\frac{2}{3}$ or $\frac{3}{8} \times \frac{2}{3}$)

$\frac{1}{4}$ of a yard

SELF-TEST 8 (Frames 1-43)

Do the following multiplications.

1. $\frac{4}{5} \times \frac{2}{3} =$
2. $\frac{1}{2} \times \frac{1}{8} =$
3. $\left(\frac{3}{4}\right)\left(\frac{11}{6}\right) =$
4. $\frac{5}{3}\left(\frac{3}{5}\right) =$
5. $\frac{9}{4} \times \frac{8}{3} =$
6. $\left(\frac{7}{5}\right)\left(\frac{12}{8}\right) =$
7. $2 \times \frac{1}{6} =$
8. $5\left(\frac{7}{10}\right) =$
9. $\frac{5}{8} \times 0 =$
10. $1 \times \frac{8}{3} =$
11. $\frac{7}{6} \times 12 =$
12. $\frac{1}{8} \times 8 =$

ANSWERS: 1. $\frac{8}{15}$ 3. $1\frac{3}{8}$ 5. 6   7. $\frac{1}{3}$ 9. 0   11. 14

2. $\frac{1}{16}$ 4. 1   6. $2\frac{1}{10}$ 8. $3\frac{1}{2}$ 10. $2\frac{2}{3}$ 12. 1

4-8 PAIRS OF RECIPROCALS

In this section, we will define what is meant by a "pair of reciprocals". The "reciprocal" concept is important because it is used to perform divisions of fractions.

44. If the product of two numbers is "1", the two numbers are called a "pair of reciprocals".

Since $7 \times \frac{1}{7} = 1$, 7 and $\frac{1}{7}$ are called a "pair of reciprocals".

We say: The reciprocal of 7 is $\frac{1}{7}$. The reciprocal of $\frac{1}{7}$ is 7.

Since $3\left(\frac{1}{3}\right) = 1$, 3 and $\frac{1}{3}$ are called a "pair of reciprocals".

We say: (a) The reciprocal of 3 is ____. (b) The reciprocal of $\frac{1}{3}$ is ____.

45. The reciprocal of any whole number is a fraction whose numerator is "1" and whose denominator is the whole number. For example:

a) $\frac{1}{3}$ b) 3

The reciprocal of 10 is $\frac{1}{10}$, since $10 \times \frac{1}{10} = 1$.

Write the reciprocal of each whole number. (a) 5 ____ (b) 13 ____ (c) 42 ____

46. The reciprocal of any fraction whose numerator is "1" is the denominator of the fraction. For example:

a) $\frac{1}{5}$ b) $\frac{1}{13}$ c) $\frac{1}{42}$

The reciprocal of $\frac{1}{8}$ is 8, since $\frac{1}{8} \times 8 = 1$.

Write the reciprocal of each fraction. (a) $\frac{1}{2}$ ____ (b) $\frac{1}{4}$ ____ (c) $\frac{1}{25}$ ____

47. Write the reciprocal of each whole number and fraction below.

a) 2 b) 4 c) 25

(a) 6 ____ (b) $\frac{1}{9}$ ____ (c) 35 ____ (d) $\frac{1}{100}$ ____

48. Since $\frac{6}{7} \times \frac{7}{6} = 1$, $\frac{6}{7}$ and $\frac{7}{6}$ are a pair of reciprocals.

a) $\frac{1}{6}$ b) 9 c) $\frac{1}{35}$ d) 100

Since $\frac{4}{3} \times \frac{3}{4} = 1$, $\frac{4}{3}$ and $\frac{3}{4}$ are a pair of _____.

reciprocals

49. Since $\left(\dfrac{5}{8}\right)\left(\dfrac{8}{5}\right) = 1$, $\dfrac{5}{8}$ and $\dfrac{8}{5}$ are a pair of reciprocals.

   (a) The reciprocal of $\dfrac{5}{8}$ is _____.   (b) The reciprocal of $\dfrac{8}{5}$ is _____.

50. To obtain the reciprocals of fractions like $\dfrac{2}{3}$ and $\dfrac{7}{4}$, we simply interchange the numerator and denominator of each. For example:

   The reciprocal of $\dfrac{2}{3}$ is $\dfrac{3}{2}$.   The reciprocal of $\dfrac{7}{4}$ is $\dfrac{4}{7}$.

   Write the reciprocal of each fraction: (a) $\dfrac{3}{5}$ _____   (b) $\dfrac{11}{9}$ _____   (c) $\dfrac{44}{57}$ _____

   a) $\dfrac{8}{5}$ b) $\dfrac{5}{8}$

51. There is one number that is its own reciprocal. That is:

   Since $1 \times 1 = 1$, the reciprocal of 1 is _____.

   a) $\dfrac{5}{3}$ b) $\dfrac{9}{11}$ c) $\dfrac{57}{44}$

52. Write the reciprocal of each number below.

   (a) $\dfrac{1}{75}$ _____   (b) 1 _____   (c) 5 _____   (d) $\dfrac{15}{4}$ _____

   1

   a) 75   b) 1   c) $\dfrac{1}{5}$ d) $\dfrac{4}{15}$

4-9 THE PROCEDURE FOR DIVIDING FRACTIONS

In this section, we will describe and use the procedure for dividing fractions. In a later section, we will show that the division procedure makes sense.

53. The names "dividend", "divisor", and "quotient" are used for the numbers in a division of whole numbers or fractions. For example:

   In $20 \div 5 = 4$: 20 is called the "dividend".
   5 is called the "divisor".
   4 is called the "quotient".

   In $\dfrac{2}{3} \div \dfrac{7}{8} = \dfrac{16}{21}$: $\dfrac{2}{3}$ is called the "dividend".

   (a) _____ is called the "divisor".

   (b) _____ is called the "quotient".

54. Any division of fractions can be converted to a multiplication of fractions by simply multiplying the dividend by the reciprocal of the divisor. For example:

   a) $\dfrac{7}{8}$ b) $\dfrac{16}{21}$

   $$\dfrac{2}{3} \div \dfrac{7}{8} = \dfrac{2}{3} \times \left(\text{the reciprocal of } \dfrac{7}{8}\right) = \dfrac{2}{3} \times \dfrac{8}{7}$$

   Complete the following conversions of a division to a multiplication.

   (a) $\dfrac{4}{3} \div \dfrac{2}{5} = \dfrac{4}{3} \times \left(\text{the reciprocal of } \dfrac{2}{5}\right) = \dfrac{4}{3} \times$ _____

   (b) $\dfrac{5}{6} \div \dfrac{9}{7} = \dfrac{5}{6} \times \left(\text{the reciprocal of } \dfrac{9}{7}\right) = \dfrac{5}{6} \times$ _____

   a) $\dfrac{4}{3} \times \dfrac{5}{2}$ b) $\dfrac{5}{6} \times \dfrac{7}{9}$

78

55. By multiplying by the reciprocal of the divisor, complete the following conversions of a division to a multiplication.

(a) $\frac{1}{4} \div \frac{3}{7} = \frac{1}{4} \times$ _____   (b) $\frac{12}{5} \div \frac{8}{3} = \left(\frac{12}{5}\right)\left(\text{———}\right)$ (c) $\frac{4}{3} \div \frac{3}{4} = \left(\frac{4}{3}\right)\left(\text{———}\right)$

56. After converting a division of fractions to a multiplication of fractions, we can find the quotient by performing the multiplication. For example:

a) $\frac{1}{4} \times \frac{7}{3}$ b) $\left(\frac{12}{5}\right)\left(\frac{3}{8}\right)$ c) $\left(\frac{4}{3}\right)\left(\frac{4}{3}\right)$

$\frac{1}{4} \div \frac{3}{5} = \frac{1}{4} \times \frac{5}{3} = \frac{5}{12}$ Therefore: $\frac{1}{4} \div \frac{3}{5} = \frac{5}{12}$

$\frac{2}{5} \div \frac{7}{9} = \frac{2}{5} \times \frac{9}{7} = \frac{18}{35}$ Therefore: $\frac{2}{5} \div \frac{7}{9} =$ _____

57. In each case below, complete the conversion of a division to a multiplication and then perform the multiplication to find the quotient.

$\frac{18}{35}$

(a) $\frac{5}{7} \div \frac{3}{2} = \frac{5}{7} \times$ _____ = _____   (b) $\frac{7}{6} \div \frac{8}{5} = \frac{7}{6} \times$ _____ = _____

58. After converting the division to a multiplication, find each quotient below. Write each quotient in lowest terms.

a) $\frac{5}{7} \times \frac{2}{3} = \frac{10}{21}$ b) $\frac{7}{6} \times \frac{5}{8} = \frac{35}{48}$

(a) $\frac{1}{2} \div \frac{5}{4} =$ _____ × _____ = _____   (b) $\frac{7}{5} \div \frac{3}{10} =$ _____ × _____ = _____

59. Complete each division below. Write each quotient in lowest terms.

a) $\frac{1}{2} \times \frac{4}{5} = \frac{2}{5}$ b) $\frac{7}{5} \times \frac{10}{3} = 4\frac{2}{3}$

(a) $\frac{9}{8} \div \frac{9}{8} = ()() =$ _____   (b) $\frac{3}{2} \div \frac{3}{8} = ()() =$ _____

60. In each case below, the reciprocal of the divisor is a whole number. Complete the bottom division.

a) $\left(\frac{9}{8}\right)\left(\frac{8}{9}\right) = 1$ b) $\left(\frac{3}{2}\right)\left(\frac{8}{3}\right) = 4$

$\frac{1}{8} \div \frac{1}{3} = \frac{1}{8} \times \left(\text{the reciprocal of } \frac{1}{3}\right) = \frac{1}{8} \times 3 = \frac{3}{8}$

$\frac{7}{10} \div \frac{1}{5} = \frac{7}{10} \times \left(\text{the reciprocal of } \frac{1}{5}\right) = \frac{7}{10} \times 5 =$ _____

61. Write each quotient in lowest terms.

$3\frac{1}{2}$

(a) $\frac{3}{7} \div \frac{1}{2} =$ _____ × _____ = _____   (b) $\frac{5}{24} \div \frac{1}{4} =$ _____ × _____ = _____

62. Write each quotient in lowest terms.

a) $\frac{3}{7} \times 2 = \frac{6}{7}$ b) $\frac{5}{24} \times 4 = \frac{5}{6}$

(a) $\frac{7}{2} \div \frac{1}{6} = ()() =$ _____   (b) $\frac{5}{16} \div \frac{1}{8} = ()() =$ _____

a) $\left(\frac{7}{2}\right)(6) = 21$ b) $\left(\frac{5}{16}\right)(8) = 2\frac{1}{2}$

4-10 DIVISIONS INVOLVING A WHOLE NUMBER AND A FRACTION

In this section, we will discuss the procedure for divisions involving a whole number and a fraction. The procedure is the same as that used for dividing fractions. In the next section, we will show that the procedure makes sense.

63. In the division below, the dividend is a whole number. We have converted the division to a multiplication by multiplying the dividend by the reciprocal of the divisor.

$$3 \div \frac{2}{5} = 3 \times \left(\text{the reciprocal of } \frac{2}{5}\right) = 3 \times \frac{5}{2}$$

Complete the following conversion of a division to a multiplication.

$$5 \div \frac{9}{7} = 5 \times \left(\text{the reciprocal of } \frac{9}{7}\right) = 5 \times \underline{}$$

64. Following the example at the left below, convert each of the other divisions to a multiplication. $\boxed{5 \times \frac{7}{9}}$

$$10 \div \frac{5}{8} = 10 \times \frac{8}{5} \qquad \text{(a)} \ 9 \div \frac{7}{2} = \underline{} \times \underline{} \qquad \text{(b)} \ 12 \div \frac{5}{6} = ()()$$

65. Convert each division below to a multiplication and then find the quotient. $\boxed{\text{a)} \ 9 \times \frac{2}{7} \quad \text{b)} \ (12)\left(\frac{6}{5}\right)}$

(a) $5 \div \frac{11}{2} = \underline{} \times \underline{} = \underline{}$ (b) $12 \div \frac{4}{3} = ()() = \underline{}$

66. When each division below was converted to a multiplication, we got a multiplication of two whole numbers. $\boxed{\text{a)} \ 5 \times \frac{2}{11} = \frac{10}{11} \quad \text{b)} \ (12)\left(\frac{3}{4}\right) = 9}$

$$9 \div \frac{1}{5} = 9 \times 5 = 45 \qquad\qquad 10 \div \frac{1}{3} = (10)(3) = 30$$

Using the same steps, find each quotient at the right.   (a) $12 \div \frac{1}{2} = \underline{} \times \underline{} = \underline{}$ (b) $20 \div \frac{1}{6} = ()() = \underline{}$

67. In the division below, the divisor is a whole number. We have converted the division to a multiplication in the usual way. $\boxed{\text{a)} \ 12 \times 2 = 24 \quad \text{b)} \ (20)(6) = 120}$

$$\frac{7}{8} \div 3 = \frac{7}{8} \times (\text{the reciprocal of } 3) = \frac{7}{8} \times \frac{1}{3}$$

Complete the following conversion of a division to a multiplication.

$$\frac{1}{2} \div 5 = \frac{1}{2} \times (\text{the reciprocal of } 5) = \frac{1}{2} \times \underline{}$$

68. Following the example at the left below, complete each of the other conversions to a multiplication. $\boxed{\frac{1}{2} \times \frac{1}{5}}$

$$\frac{3}{7} \div 9 = \frac{3}{7} \times \frac{1}{9} \qquad \text{(a)} \ \frac{9}{4} \div 8 = \underline{} \times \underline{} \qquad \text{(b)} \ \frac{1}{9} \div 3 = ()()$$

$\boxed{\text{a)} \ \frac{9}{4} \times \frac{1}{8} \quad \text{b)} \ \left(\frac{1}{9}\right)\left(\frac{1}{3}\right)}$

80

69. Convert each division below to a multiplication and then find the quotient.

(a) $\frac{3}{4} \div 2 =$ _____ x _____ = _____ (b) $\frac{8}{5} \div 4 =$ $\left(\right)\left(\right) =$ _____

70. Complete: (a) $6 \div \frac{9}{5} =$ _____

(b) $\frac{2}{3} \div 6 =$ _____

a) $\frac{3}{4} \times \frac{1}{2} = \frac{3}{8}$ b) $\left(\frac{8}{5}\right)\left(\frac{1}{4}\right) = \frac{2}{5}$

71. Complete: (a) $7 \div \frac{1}{4} =$ _____ (b) $\frac{5}{2} \div 10 =$ _____

a) $3\frac{1}{3}$ b) $\frac{1}{9}$

a) 28 b) $\frac{1}{4}$

4-11 JUSTIFYING THE PROCEDURE FOR DIVISIONS INVOLVING FRACTIONS

In the last two sections, we described and used the procedure for divisions involving fractions. In this section, we will justify the procedure by showing that it makes sense. To do so, we will use the relationship between multiplication and division.

72. When discussing operations with whole numbers, we showed that there is a relationship between division and multiplication. At that time, we showed this fact: <u>If a division is performed correctly, the product of the divisor and the quotient equals the dividend</u>. For example:

$$21 \div 3 = 7 \quad \text{and} \quad 3 \times 7 = 21$$

If the procedure we are using to perform divisions involving fractions makes sense, the same fact should be true. That is, the product of the divisor and the quotient should equal the dividend. Let's check this fact for the division below.

$$\frac{1}{2} \div \frac{4}{3} = \frac{3}{8} \quad \left(\text{from } \frac{1}{2} \div \frac{4}{3} = \frac{1}{2} \times \frac{3}{4} = \frac{3}{8}\right)$$

(a) At the right, multiply $\frac{4}{3}$ and $\frac{3}{8}$ to see whether the product of the divisor and the quotient equals the dividend "$\frac{1}{2}$".

$\frac{4}{3} \times \frac{3}{8} =$ _____

(b) Therefore, does the division procedure make sense? _____

73. Let's check to see whether the same fact holds for the division below in which the quotient is a whole number.

a) $\frac{1}{2}$ b) Yes

$$\frac{1}{3} \div \frac{1}{6} = 2 \quad \left(\text{from } \frac{1}{3} \div \frac{1}{6} = \frac{1}{3} \times 6 = 2\right)$$

(a) At the right, multiply $\frac{1}{6}$ and 2 to see whether the product of the divisor and the quotient equals the dividend "$\frac{1}{3}$".

$\frac{1}{6} \times 2 =$ _____

(b) Therefore, does the division procedure make sense? _____

a) $\frac{1}{3}$ b) Yes

74. Let's check to see whether the same fact holds for the division below in which the dividend is a whole number.

$$2 \div \frac{7}{3} = \frac{6}{7} \quad \left(\text{from } 2 \div \frac{7}{3} = 2 \times \frac{3}{7} = \frac{6}{7}\right)$$

(a) At the right, multiply $\frac{7}{3}$ and $\frac{6}{7}$ to see whether their product equals the dividend "2".

$$\frac{7}{3} \times \frac{6}{7} = \underline{}$$

(b) Therefore, does the division procedure make sense? _____

---

a) 2   b) Yes

75. Let's check to see whether the same fact holds for the division below in which the divisor is a whole number.

$$\frac{1}{4} \div 3 = \frac{1}{12} \quad \left(\text{from } \frac{1}{4} \div 3 = \frac{1}{4} \times \frac{1}{3} = \frac{1}{12}\right)$$

(a) At the right, multiply 3 and $\frac{1}{12}$ to see whether their product equals the dividend $\frac{1}{4}$.

$$3 \times \frac{1}{12} = \underline{}$$

(b) Therefore, does the division procedure make sense? _____

---

a) $\frac{1}{4}$ b) Yes

76. Let's check to see whether the same fact holds for the division below in which both the dividend and the quotient are whole numbers.

$$5 \div \frac{1}{4} = 20 \quad \left(\text{from } 5 \div \frac{1}{4} = 5 \times 4 = 20\right)$$

(a) At the right, multiply $\frac{1}{4}$ and 20 to see whether their product equals the dividend "5".

$$\frac{1}{4} \times 20 = \underline{}$$

(b) Therefore, does the division procedure make sense? _____

a) 5   b) Yes

4-12   DIVISION OF FRACTIONS IN COMPLEX-FRACTION FORM

Up to this point, we have used the symbol "÷" as the division symbol for divisions involving fractions. But just as divisions of whole numbers can be written as fractions, divisions involving fractions can be written as fractions. We will discuss the complex-fraction form of divisions involving fractions in this section.

77. We saw earlier that any division of whole numbers can be written as a fraction. For example:

$20 \div 4$ can be written $\frac{20}{4}$ $17 \div 8$ can be written $\frac{17}{8}$

Any division of fractions can also be written as a single fraction. Since more than one fraction line is involved, the single fraction is called a "complex" fraction. For example:

$\frac{3}{4} \div \frac{5}{8}$ can be written $\dfrac{\frac{3}{4}}{\frac{5}{8}}$ ⟵ $\frac{5}{12} \div \frac{1}{3}$ can be written $\dfrac{\frac{5}{12}}{\frac{1}{3}}$ ⟵

Note: In each case, <u>one of the fraction lines is longer</u>. (See the arrows.) This longer fraction line is called the "<u>major</u>" fraction line.

Continued on following page.

77. Continued

Write each division at the right as a complex fraction.

(a) $\frac{7}{3} \div \frac{1}{4} =$ _____

(b) $\frac{1}{2} \div \frac{9}{10} =$ _____

---

78. In the complex fraction $\dfrac{\frac{1}{2}}{\frac{5}{7}}$: the numerator is $\frac{1}{2}$ and the denominator is $\frac{5}{7}$.

a) $\dfrac{\frac{7}{3}}{\frac{1}{4}}$ b) $\dfrac{\frac{1}{2}}{\frac{9}{10}}$

In the complex fraction $\dfrac{\frac{5}{6}}{\frac{7}{3}}$: (a) the numerator is _____. (b) the denominator is _____.

---

79. Any complex fraction also stands for a division in which the numerator is divided by the denominator. That is:

a) $\frac{5}{6}$ b) $\frac{7}{3}$

$\dfrac{\frac{2}{5}}{\frac{3}{8}}$ means "divide $\frac{2}{5}$ by $\frac{3}{8}$".

$\dfrac{\frac{1}{3}}{\frac{1}{7}}$ means "divide _____ by _____".

---

80. Any division involving a whole number and a fraction can also be written as a complex fraction. For example:

$\frac{1}{3}$ by $\frac{1}{7}$

$\frac{5}{9} \div 4$ can be written $\dfrac{\frac{5}{9}}{4}$ ↙

$3 \div \frac{1}{6}$ can be written $\dfrac{3}{\frac{1}{6}}$ ↙

Note: In each case, the "major" fraction line is longer. (See the arrows.)

Write each division at the right as a complex fraction.

(a) $\frac{7}{11} \div 9 =$ _____

(b) $1 \div \frac{3}{4} =$ _____

---

81. In the complex fraction $\dfrac{\frac{7}{10}}{4}$: the numerator is $\frac{7}{10}$ and the denominator is 4.

a) $\dfrac{\frac{7}{11}}{9}$ b) $\dfrac{1}{\frac{3}{4}}$

In the complex fraction $\dfrac{1}{\frac{1}{5}}$: (a) the numerator is _____.
(b) the denominator is _____.

---

82. Complex fractions involving a whole number and a fraction also stand for divisions. That is:

a) 1   b) $\frac{1}{5}$

$\dfrac{\frac{3}{4}}{10}$ means "divide $\frac{3}{4}$ by 10".

$\dfrac{5}{\frac{2}{9}}$ means "divide _____ by _____".

---

5 by $\frac{2}{9}$

83. When a division is written in complex-fraction form, we can also convert the division to a multiplication. To do so, we multiply the numerator by the reciprocal of the denominator. That is:

$$\frac{\frac{5}{9}}{\frac{4}{11}} = \frac{5}{9} \times \left(\text{the reciprocal of } \frac{4}{11}\right) = \frac{5}{9} \times \frac{11}{4}$$

Convert each division to a multiplication.

(a) $\dfrac{\frac{3}{7}}{\frac{4}{9}} = $ _____ x _____

(b) $\dfrac{\frac{11}{17}}{\frac{47}{5}} = ($ _____ $)($ _____ $)$

---

84. The division below has been converted to a multiplication by multiplying the numerator by the reciprocal of the denominator.

$$\frac{7}{\frac{8}{5}} = 7 \times \left(\text{the reciprocal of } \frac{8}{5}\right) = 7 \times \frac{5}{8}$$

a) $\dfrac{3}{7} \times \dfrac{9}{4}$ b) $\left(\dfrac{11}{17}\right)\left(\dfrac{5}{47}\right)$

Convert each division to a multiplication.

(a) $\dfrac{9}{\frac{2}{11}} = $ _____ x _____

(b) $\dfrac{4}{\frac{1}{6}} = ($ _____ $)($ _____ $)$

---

85. The division below has been converted to a multiplication by multiplying the numerator by the reciprocal of the denominator.

$$\frac{\frac{4}{5}}{6} = \frac{4}{5} \times (\text{the reciprocal of } 6) = \frac{4}{5} \times \frac{1}{6}$$

a) $9 \times \dfrac{11}{2}$ b) $(4)(6)$

Convert each division to a multiplication.

(a) $\dfrac{\frac{2}{3}}{7} = $ _____ x _____

(b) $\dfrac{\frac{1}{5}}{3} = ($ _____ $)($ _____ $)$

---

86. Having converted a division in complex-fraction form to a multiplication, we can find the quotient by performing the multiplication. Find each quotient below.

a) $\dfrac{2}{3} \times \dfrac{1}{7}$ b) $\left(\dfrac{1}{5}\right)\left(\dfrac{1}{3}\right)$

(a) $\dfrac{\frac{5}{7}}{\frac{3}{2}} = $ _____ x _____ = _____

(b) $\dfrac{\frac{1}{4}}{\frac{2}{7}} = $ _____ x _____ = _____

---

87. Find each quotient below.

a) $\dfrac{5}{7} \times \dfrac{2}{3} = \dfrac{10}{21}$ b) $\dfrac{1}{4} \times \dfrac{7}{2} = \dfrac{7}{8}$

(a) $\dfrac{3}{\frac{8}{5}} = $ _____ x _____ = _____

(b) $\dfrac{7}{\frac{1}{8}} = ($ _____ $)($ _____ $) = $ _____

a) $3 \times \dfrac{5}{8} = 1\dfrac{7}{8}$ b) $(7)(8) = 56$

88. Find each quotient below.

(a) $\dfrac{\frac{2}{9}}{3} = \underline{\hspace{1cm}} \times \underline{\hspace{1cm}} = \underline{\hspace{1cm}}$

(b) $\dfrac{\frac{1}{6}}{7} = \left(\underline{\hspace{1cm}}\right)\left(\underline{\hspace{1cm}}\right) = \underline{\hspace{1cm}}$

a) $\dfrac{2}{9} \times \dfrac{1}{3} = \dfrac{2}{27}$ b) $\left(\dfrac{1}{6}\right)\left(\dfrac{1}{7}\right) = \dfrac{1}{42}$

4-13 DIVISIONS INVOLVING A FRACTION AND EITHER "1" OR "0"

In this section, we will briefly discuss divisions involving a fraction and either "1" or "0". Divisions of a fraction by itself are also discussed.

89. <u>Whenever we divide a fraction by itself, the quotient is "1"</u>. For example:

$$\dfrac{2}{3} \div \dfrac{2}{3} = \dfrac{2}{3} \times \dfrac{3}{2} = 1 \qquad\qquad \dfrac{\frac{7}{4}}{\frac{7}{4}} = \dfrac{7}{4} \times \dfrac{4}{7} = 1$$

Using the fact above, write each quotient at the right.

(a) $\dfrac{9}{5} \div \dfrac{9}{5} = \underline{\hspace{1cm}}$ (b) $\dfrac{\frac{1}{2}}{\frac{1}{2}} = \underline{\hspace{1cm}}$

90. <u>Whenever we divide a fraction by "1", the quotient is the fraction</u>. For example:

a) 1   b) 1

$$\dfrac{2}{5} \div 1 = \dfrac{2}{5} \times 1 = \dfrac{2}{5} \qquad\qquad \dfrac{\frac{9}{7}}{1} = \dfrac{9}{7} \times 1 = \dfrac{9}{7} = 1\dfrac{2}{7}$$

Note: The reciprocal of "1" is "1", since 1 x 1 = 1.

Using the fact above, write each quotient at the right.

(a) $\dfrac{8}{3} \div 1 = \underline{\hspace{1cm}}$ (b) $\dfrac{\frac{4}{9}}{1} = \underline{\hspace{1cm}}$

91. <u>Whenever we divide "0" by a fraction, the quotient is "0"</u>. For example:

a) $2\dfrac{2}{3}$ (from $\dfrac{8}{3}$)   b) $\dfrac{4}{9}$

$$0 \div \dfrac{3}{4} = 0 \times \dfrac{4}{3} = 0 \qquad\qquad \dfrac{0}{\frac{1}{8}} = 0 \times 8 = 0$$

Using the fact above, write each quotient at the right.

(a) $0 \div \dfrac{1}{3} = \underline{\hspace{1cm}}$ (b) $\dfrac{0}{\frac{5}{2}} = \underline{\hspace{1cm}}$

92. Complete: (a) $\dfrac{1}{4} \div \dfrac{1}{4} = \underline{\hspace{1cm}}$ (b) $\dfrac{1}{4} \div 1 = \underline{\hspace{1cm}}$ (c) $0 \div \dfrac{1}{4} = \underline{\hspace{1cm}}$

a) 0   b) 0

a) 1   b) $\dfrac{1}{4}$ c) 0

93. Complete: (a) $\dfrac{\frac{7}{8}}{\frac{7}{8}} = \underline{\hspace{1cm}}$ (b) $\dfrac{\frac{0}{7}}{\frac{7}{8}} = \underline{\hspace{1cm}}$ (c) $\dfrac{\frac{7}{8}}{1} = \underline{\hspace{1cm}}$

94. In each case below, we have divided "1" by a fraction.

    $1 \div \dfrac{5}{6} = 1 \times \dfrac{6}{5} = \dfrac{6}{5} = 1\dfrac{1}{5}$ $\dfrac{1}{\frac{7}{2}} = 1 \times \dfrac{2}{7} = \dfrac{2}{7}$

    | a) 1   b) 0   c) $\dfrac{7}{8}$ |

    Following the same steps, find each quotient at the right.
    (a) $1 \div \dfrac{5}{3} = 1 \times \underline{\hspace{1cm}} = \underline{\hspace{1cm}}$ (b) $\dfrac{1}{\frac{1}{9}} = 1 \times \underline{\hspace{1cm}} = \underline{\hspace{1cm}}$

95. Complete:

    | a) $1 \times \dfrac{3}{5} = \dfrac{3}{5}$ b) $1 \times 9 = 9$ |

    (a) $1 \div \dfrac{3}{7} = \underline{\hspace{1cm}}$ (b) $\dfrac{3}{7} \div 1 = \underline{\hspace{1cm}}$ (c) $\dfrac{3}{7} \div \dfrac{3}{7} = \underline{\hspace{1cm}}$

96. Complete:

    | a) $2\dfrac{1}{3}$ (from $\dfrac{7}{3}$)   b) $\dfrac{3}{7}$ c) 1 |

    (a) $\dfrac{\frac{1}{5}}{1} = \underline{\hspace{1cm}}$ (b) $\dfrac{0}{\frac{1}{5}} = \underline{\hspace{1cm}}$ (c) $\dfrac{1}{\frac{1}{5}} = \underline{\hspace{1cm}}$

    | a) $\dfrac{1}{5}$ b) 0   c) 5 |

4-14 APPLIED PROBLEMS

This section contains some verbal or applied problems. All of them can be solved by performing a division involving a fraction.

97. A man buys $\dfrac{3}{4}$ of an acre of land and divides it into three equal lots. How large is each lot?

    (Think: $\dfrac{3}{4}$ divided by 3)

    | $\dfrac{1}{4}$ of an acre |

98. If $\dfrac{5}{8}$ of a mile is divided into 10 equal parts, how long is each part?

    (Think: $\dfrac{5}{8}$ divided by 10)

    | $\dfrac{1}{16}$ of a mile |

99. A store buys a large bottle of perfume containing 48 ounces. It sells the perfume in small bottles that contain $\dfrac{3}{8}$ of an ounce. How many of the small bottles can be filled?

    (Think: 48 divided by $\dfrac{3}{8}$)

    | 128 bottles |

100. A roll of wire is 36 inches long. How many pieces that are $\dfrac{9}{16}$ of an inch long can be cut from the roll?   (Think: 36 divided by $\dfrac{9}{16}$)

    | 64 pieces |

SELF-TEST 9 (Frames 44-100)

1. $\frac{5}{12} \div \frac{1}{2} =$

2. $\frac{9}{8} \div \frac{3}{10} =$

3. $\frac{5}{2} \div 1 =$

4. $1 \div \frac{1}{4} =$

5. $0 \div \frac{7}{8} =$

6. $\frac{3}{5} \div \frac{3}{5} =$

7. $\frac{2}{3} \div \frac{3}{2} =$

8. $12 \div \frac{4}{5} =$

9. $\frac{4}{3} \div 8 =$

10. $\dfrac{\frac{3}{2}}{\frac{9}{4}} =$

11. $\dfrac{8}{\frac{6}{5}} =$

12. $\dfrac{\frac{3}{4}}{6} =$

13. $\dfrac{\frac{1}{7}}{2} =$

ANSWERS:

1. $\frac{5}{6}$
2. $3\frac{3}{4}$
3. $2\frac{1}{2}$
4. 4
5. 0
6. 1
7. $\frac{4}{9}$
8. 15
9. $\frac{1}{6}$
10. $\frac{2}{3}$
11. $6\frac{2}{3}$
12. $\frac{1}{8}$
13. $\frac{2}{7}$

4-15 MULTIPLYING THREE FRACTIONS

In this section, we will discuss the procedure for multiplications involving three fractions.

---

101. The multiplication at the right was performed by multiplying "two at a time" from left to right. That is:

   (1) We began by multiplying $\frac{1}{2}$ and $\frac{3}{4}$ and got $\frac{3}{8}$.

   (2) Then we multiplied $\frac{3}{8}$ and $\frac{5}{7}$ to get the final product $\frac{15}{56}$.

$\frac{1}{2} \times \frac{3}{4} \times \frac{5}{7}$

$\frac{3}{8} \times \frac{5}{7} = \frac{15}{56}$

Following the steps above, complete each multiplication at the right.

(a) $\underline{\frac{1}{5} \times \frac{3}{2}} \times \frac{9}{4}$

_____ $\times \frac{9}{4} =$ _____

(b) $\underline{\frac{1}{3} \times \frac{1}{2}} \times \frac{1}{6}$

_____ $\times \frac{1}{6} =$ _____

102. The multiplication at the right was performed in the last frame. The final product is $\frac{15}{56}$.

We can obtain the product for the same multiplication by simply multiplying the three numerators and multiplying the three denominators. That is:

$$\frac{1}{2} \times \frac{3}{4} \times \frac{5}{7} = \frac{1 \times 3 \times 5}{2 \times 4 \times 7} = \frac{15}{56}$$

$\frac{1}{2} \times \frac{3}{4} \times \frac{5}{7}$

$\frac{3}{8} \times \frac{5}{7} = \frac{15}{56}$

a) $\frac{3}{10} \times \frac{9}{4} = \frac{27}{40}$

b) $\frac{1}{6} \times \frac{1}{6} = \frac{1}{36}$

Continued on following page.

102. Continued

By simply multiplying the three numerators and multiplying the three denominators, find each product.

(a) $\frac{3}{2} \times \frac{1}{4} \times \frac{7}{5} =$ _____ (b) $\frac{1}{3} \times \frac{1}{2} \times \frac{1}{4} =$ _____

103. Parentheses can also be used as the multiplication symbol in multiplications of three fractions. For example:

Both $\left(\frac{5}{6}\right)\left(\frac{1}{4}\right)\left(\frac{7}{2}\right)$ and $\frac{5}{6}\left(\frac{1}{4}\right)\left(\frac{7}{2}\right)$ mean "$\frac{5}{6} \times \frac{1}{4} \times \frac{7}{2}$"

Find each product. (a) $\left(\frac{1}{2}\right)\left(\frac{7}{5}\right)\left(\frac{1}{2}\right) =$ _____ (b) $\frac{5}{4}\left(\frac{1}{3}\right)\left(\frac{1}{2}\right) =$ _____

a) $\frac{21}{40}$ b) $\frac{1}{24}$

104. To show that the "order" principle of multiplication also applies to multiplications of three fractions, we have written the same three fractions in various orders below. The three multiplications are equal since the product for each is _____ .

$\frac{1}{2} \times \frac{3}{4} \times \frac{7}{5}$ $\frac{3}{4} \times \frac{7}{5} \times \frac{1}{2}$ $\frac{7}{5} \times \frac{1}{2} \times \frac{3}{4}$

a) $\frac{7}{20}$ b) $\frac{5}{24}$

$\frac{21}{40}$

4-16 THE "CANCELLING" PROCESS FOR MULTIPLICATIONS INVOLVING THREE FRACTIONS

The "cancelling" process can also be used to simplify some multiplications involving three fractions. We will discuss this use of the "cancelling" process in this section.

105. We found the product at the right in the usual way by simply multiplying the numerators and denominators. Notice that the product had to be reduced to lowest terms.

$\frac{8}{5} \times \frac{7}{3} \times \frac{1}{4} = \frac{56}{60} = \frac{14}{15}$

The same multiplication was simplified by using the "cancelling" process at the right. That is, we divided both the "8" and the "4" by 4 before multiplying.

$\frac{\overset{2}{\cancel{8}}}{5} \times \frac{7}{3} \times \frac{1}{\underset{1}{\cancel{4}}} = \frac{14}{15}$

Before performing the multiplication at the right, "cancel" by dividing both the "10" and the "5" by 5.

$\frac{3}{10} \times \frac{1}{4} \times \frac{5}{2} =$ _____

106. We used cancelling to simplify the multiplication at the right. That is, before multiplying, we divided the "3" and "6" by 3 and the "7" and "7" by 7.

$\left(\frac{\overset{1}{\cancel{3}}}{\cancel{7}}\right)\left(\frac{\overset{1}{\cancel{7}}}{\cancel{6}}\right)\left(\frac{5}{8}\right) = \frac{5}{16}$
$1 2$

$\frac{3}{\underset{2}{\cancel{10}}} \times \frac{1}{4} \times \frac{\overset{1}{\cancel{5}}}{2} = \frac{3}{16}$

Before performing the multiplication at the right, "cancel" by dividing the "16" and "8" by 8 and the "5" and "15" by 5.

$\left(\frac{3}{7}\right)\left(\frac{16}{5}\right)\left(\frac{15}{8}\right) =$ _____

$2\frac{4}{7}$, from:

$\left(\frac{3}{7}\right)\left(\frac{\overset{2}{\cancel{16}}}{\cancel{5}}\right)\left(\frac{\overset{3}{\cancel{15}}}{\cancel{8}}\right) = \frac{18}{7}$
$1 1$

87

107. By cancelling, we were able to simplify all of the terms at the right. That is, we divided the "4" and "8" by 4, the "5" and "20" by 5, and the "9" and "21" by 3.

$$\frac{\overset{1}{\cancel{4}}}{\underset{3}{\cancel{9}}} \times \frac{\overset{1}{\cancel{5}}}{\underset{2}{\cancel{8}}} \times \frac{\overset{7}{\cancel{21}}}{\underset{4}{\cancel{20}}} = \frac{7}{24}$$

By cancelling, you can simplify all of the terms at the right before multiplying. Do so.

$$\frac{9}{10} \times \frac{49}{6} \times \frac{5}{7} = ____$$

$5\frac{1}{4}$, from:

$$\frac{\overset{3}{\cancel{9}}}{\underset{2}{\cancel{10}}} \times \frac{\overset{7}{\cancel{49}}}{\underset{2}{\cancel{6}}} \times \frac{\overset{1}{\cancel{5}}}{\underset{1}{\cancel{7}}} = \frac{21}{4}$$

108. We used cancelling to simplify each multiplication below.

$$\frac{\overset{1}{\cancel{4}}}{\underset{5}{\cancel{25}}} \times \frac{\overset{1}{\cancel{3}}}{\underset{2}{\cancel{8}}} \times \frac{\overset{1}{\cancel{5}}}{\underset{3}{\cancel{9}}} = \frac{1}{30} \qquad \left(\frac{\overset{1}{\cancel{7}}}{\underset{1}{\cancel{4}}}\right)\left(\frac{\overset{2}{\cancel{8}}}{\underset{1}{\cancel{3}}}\right)\left(\frac{\overset{3}{15}}{\underset{1}{\cancel{7}}}\right) = \frac{6}{1} = 6$$

Note: (1) In the multiplication <u>on the left</u>, the numerator of the product is "1". It was obtained from the multiplication 1 x 1 x 1 .

(2) In the multiplication <u>on the right</u>, the denominator of the product is "1". Therefore, the product is written as a whole number.

Use cancelling to simplify each multiplication at the right.

(a) $\left(\frac{10}{3}\right)\left(\frac{12}{11}\right)\left(\frac{11}{5}\right) = ____$ (b) $\frac{17}{4} \times \frac{2}{15} \times \frac{3}{17} = ____$

---

a) 8, from:

$$\left(\frac{\overset{2}{\cancel{10}}}{\underset{1}{\cancel{3}}}\right)\left(\frac{\overset{4}{\cancel{12}}}{\underset{1}{11}}\right)\left(\frac{\overset{1}{\cancel{11}}}{\underset{1}{\cancel{5}}}\right) = \frac{8}{1}$$

b) $\frac{1}{10}$, from:

$$\frac{\overset{1}{\cancel{17}}}{\underset{2}{\cancel{4}}} \times \frac{\overset{1}{\cancel{2}}}{\underset{5}{\cancel{15}}} \times \frac{\overset{1}{\cancel{3}}}{\underset{1}{\cancel{17}}} = \frac{1}{10}$$

109. We were able to do a "double" cancelling with the numerator "6" at the right. That is:

$$\frac{\overset{\overset{1}{\cancel{3}}}{\cancel{6}}}{5} \times \frac{1}{\underset{2}{\cancel{4}}} \times \frac{7}{\underset{1}{\cancel{3}}} = \frac{7}{10}$$

(1) First we divided the "6" and "4" by 2.

(2) Then we divided the remaining 3 and the denominator "3" by 3.

A "double" cancelling is possible with the numerator "20" at the right. Cancel and find the product.

$$\frac{20}{7} \times \frac{1}{2} \times \frac{3}{5} = ____$$

$\frac{6}{7}$, from:

$$\frac{\overset{\overset{2}{\cancel{10}}}{\cancel{20}}}{7} \times \frac{1}{\underset{1}{\cancel{2}}} \times \frac{3}{\underset{1}{\cancel{5}}} = \frac{6}{7}$$

110. We were able to do a "double" cancelling with the denominator "18" at the right. That is:

$$\left(\frac{\overset{1}{\cancel{3}}}{7}\right)\left(\frac{11}{\underset{\underset{1}{\cancel{6}}}{\cancel{18}}}\right)\left(\frac{\overset{2}{\cancel{12}}}{5}\right) = \frac{22}{35}$$

(1) First we divided the "3" and "18" by 3.

(2) Then we divided the remaining 6 and the numerator "12" by 6.

A "double" cancelling is possible with the denominator "12" at the right. Cancel and find the product.

$$\left(\frac{7}{12}\right)\left(\frac{4}{5}\right)\left(\frac{3}{7}\right) = ____$$

$\frac{1}{5}$, from:

$$\left(\frac{\overset{1}{\cancel{7}}}{\underset{\underset{1}{\cancel{3}}}{\cancel{12}}}\right)\left(\frac{\overset{1}{\cancel{4}}}{5}\right)\left(\frac{\overset{1}{\cancel{3}}}{\underset{1}{\cancel{7}}}\right) = \frac{1}{5}$$

4-17 THREE-FACTOR MULTIPLICATIONS INVOLVING FRACTIONS AND WHOLE NUMBERS

In this section, we will discuss the procedure for performing three-factor multiplications involving both fractions and whole numbers.

111. Any three-factor multiplication involving one whole number and two fractions can be converted to a multiplication of three fractions by substituting a fraction for the whole number. For example:

$$3 \times \frac{1}{4} \times \frac{5}{7} = \frac{3}{1} \times \frac{1}{4} \times \frac{5}{7} = \frac{15}{28} \qquad \left(\frac{4}{5}\right)(2)\left(\frac{1}{3}\right) = \left(\frac{4}{5}\right)\left(\frac{2}{1}\right)\left(\frac{1}{3}\right) = \frac{8}{15}$$

Note: We substituted $\frac{3}{1}$ for 3 on the left and $\frac{2}{1}$ for 2 on the right.

Substitute $\frac{4}{1}$ for 4 at the right and then find the product.

$$\frac{6}{7} \times \frac{1}{5} \times 4 = \frac{6}{7} \times \frac{1}{5} \times \underline{\qquad} = \underline{\qquad}$$

$$\boxed{\frac{6}{7} \times \frac{1}{5} \times \frac{4}{1} = \frac{24}{35}}$$

112. Any three-factor multiplication involving two whole numbers and a fraction can be converted to a multiplication of three fractions by substituting fractions for the whole numbers. Two examples are given below.

$$2 \times 3 \times \frac{1}{7} = \frac{2}{1} \times \frac{3}{1} \times \frac{1}{7} = \frac{6}{7} \qquad (6)\left(\frac{1}{5}\right)(3) = \left(\frac{6}{1}\right)\left(\frac{1}{5}\right)\left(\frac{3}{1}\right) = \frac{18}{5} = 3\frac{3}{5}$$

Substitute a fraction for each whole number at the right and then find the product.

$$\frac{1}{3} \times 4 \times 2 = \frac{1}{3} \times \underline{\qquad} \times \underline{\qquad} = \underline{\qquad}$$

$$\boxed{\frac{1}{3} \times \frac{4}{1} \times \frac{2}{1} = \frac{8}{3} = 2\frac{2}{3}}$$

113. The fraction $\frac{1}{1}$ can be substituted for the number "1". For example:

$$1 \times \frac{1}{4} \times \frac{1}{2} = \frac{1}{1} \times \frac{1}{4} \times \frac{1}{2} = \frac{1}{8} \qquad (4)\left(\frac{1}{7}\right)(1) = \left(\frac{4}{1}\right)\left(\frac{1}{7}\right)\left(\frac{1}{1}\right) = \underline{\qquad}$$

$$\boxed{\frac{4}{7}}$$

114. The multiplications below were performed by substituting a fraction for each whole number.

$$2 \times \frac{1}{3} \times \frac{10}{9} = \frac{2}{1} \times \frac{1}{3} \times \frac{10}{9} = \frac{20}{27} \qquad (3)(2)\frac{1}{7} = \left(\frac{3}{1}\right)\left(\frac{2}{1}\right)\left(\frac{1}{7}\right) = \frac{6}{7}$$

There is a shorter method that can be used for the same multiplications. In the shorter method, we simply treat each whole number like a numerator. That is:

$$2 \times \frac{1}{3} \times \frac{10}{9} = \frac{2 \times 1 \times 10}{3 \times 9} = \frac{20}{27} \qquad (3)(2)\left(\frac{1}{7}\right) = \frac{3 \times 2 \times 1}{7} = \frac{6}{7}$$

Using the shorter method, complete each multiplication below.

(a) $\frac{2}{5} \times 3 \times \frac{2}{7} = \frac{2 \times 3 \times 2}{5 \times 7} = \underline{\qquad}$

(b) $1 \times \frac{7}{3} \times 4 = \frac{1 \times 7 \times 4}{3} = \underline{\qquad}$

115. Use the shorter method for each of these:

(a) $\left(\frac{1}{4}\right)\left(\frac{3}{4}\right)(5) = \underline{\qquad}$

(b) $\left(\frac{7}{8}\right)(1)(3) = \underline{\qquad}$

a) $\frac{12}{35}$

b) $9\frac{1}{3}$ (from $\frac{28}{3}$)

116. In the multiplication at the left below, we had to reduce the product to lowest terms. Notice how we simplified the same multiplication at the bottom by "cancelling".

$$\frac{3}{4} \times 2 \times 5 = \frac{3 \times 2 \times 5}{4} = \frac{30}{4} = 7\frac{2}{4} = 7\frac{1}{2}$$

$$\frac{3}{\cancel{4}_2} \times \cancel{2}^1 \times 5 = \frac{3 \times 1 \times 5}{2} = \frac{15}{2} = 7\frac{1}{2}$$

a) $\frac{15}{16}$, from: $\frac{1 \times 3 \times 5}{4 \times 4}$

b) $2\frac{5}{8}$, from: $\frac{7 \times 1 \times 3}{8}$

Use cancelling to simplify each multiplication at the right.

(a) $3 \times \frac{8}{9} \times 1 = $ _____

(b) $(10)(5)\left(\frac{1}{6}\right) = $ _____

---

117. We used cancelling to simplify the multiplication at the left below. Since the denominator of the product is "1", the product equals a whole number. Complete the other multiplication.

$$\frac{1}{\cancel{4}_1} \times \cancel{8}^2 \times 3 = \frac{6}{1} = 6 \qquad (3)(30)\left(\frac{1}{5}\right) = $$ _____

a) $2\frac{2}{3}$, from: $\cancel{3}^1 \times \frac{8}{\cancel{9}_3} \times 1 = \frac{8}{3}$

b) $8\frac{1}{3}$, from: $(\cancel{10})^5 (5)\left(\frac{1}{\cancel{6}_3}\right) = \frac{25}{3}$

---

118. Before performing the multiplication at the left below, we cancelled by dividing the "3" and "9" by 3 and the "10" and "10" by 10. After cancelling, find the other product.

$$\frac{1}{\cancel{3}} \times \frac{7}{\cancel{10}_1} \times \frac{\cancel{10}^1}{\cancel{9}_3} = \frac{7}{3} = 2\frac{1}{3} \qquad \frac{9}{11} \times 11 \times \frac{5}{6} = $$ _____

18, from: $(3)(\cancel{30})^6 \left(\frac{1}{\cancel{5}_1}\right) = \frac{18}{1}$

---

119. A "double" cancelling was possible with the whole number "18" at the left below. Find the other product.

$$\cancel{18}^3 \left(\frac{5}{\cancel{8}_4}\right)\left(\frac{1}{\cancel{3}_1}\right) = \frac{15}{4} = 3\frac{3}{4} \qquad \frac{3}{25} \times 35 \times \frac{1}{7} = $$ _____

$7\frac{1}{2}$, from: $\frac{\cancel{9}^3}{\cancel{11}_1} \times \cancel{11}^1 \times \frac{5}{\cancel{6}_2} = \frac{15}{2}$

---

120. The "order" principle also applies to three-factor multiplications involving both whole numbers and fractions. That is, we can write the factors in any order without changing the final product. For example:

$\frac{1}{5} \times 2 \times \frac{3}{7}$ and $2 \times \frac{1}{5} \times \frac{3}{7}$ are equal since both equal $\frac{6}{35}$.

$\frac{2}{7} \times 4 \times 3$ and $4 \times 3 \times \frac{2}{7}$ are equal since both equal _____.

$\frac{3}{5}$, from: $\frac{3}{\cancel{25}_5} \times \cancel{35}^{\cancel{7}^1} \times \frac{1}{\cancel{7}_1}$

$3\frac{3}{7}$ (from $\frac{24}{7}$)

4-18 CONTRASTING THE FOUR OPERATIONS WITH FRACTIONS

The four basic operations with fractions are addition, subtraction, multiplication, and division. We will contrast those four operations in this section.

---

121. The only two operations that require common denominators are <u>addition</u> and <u>subtraction</u>. For example:

$$\frac{1}{2} + \frac{1}{4} = \frac{2}{4} + \frac{1}{4} = \frac{3}{4} \qquad\qquad \frac{1}{2} - \frac{1}{4} = \frac{2}{4} - \frac{1}{4} = \underline{\quad}$$

---

122. There is a tendency to confuse <u>addition</u> and <u>multiplication</u> of fractions. Remember that multiplication does not require common denominators since it is performed by simply multiplying the numerators and denominators as they stand. For example:

$$\frac{1}{2} + \frac{1}{3} = \frac{3}{6} + \frac{2}{6} = \frac{5}{6}, \text{ but } \frac{1}{2} \times \frac{1}{3} = \frac{1 \times 1}{2 \times 3} = \underline{\quad}$$

$\frac{1}{4}$

---

123. Ordinarily, division of fractions is not confused with the other operations. The main problem with division is remembering what to do. Remember that division is performed <u>by multiplying</u> <u>the dividend by the reciprocal of the divisor</u>. That is:

$\frac{1}{2} \div \frac{5}{6}$ is performed by multiplying $\frac{1}{2}$ by $\frac{6}{5}$. $\qquad \frac{3}{4} \div \frac{1}{7}$ is performed by multiplying $\frac{3}{4}$ by $\underline{\quad}$.

$\frac{1}{6}$

---

124. In the following few frames, we will give some mixed practice involving all four operations.

Complete: (a) $\frac{1}{3} + \frac{1}{4} =$ ____ (b) $\frac{1}{3} \times \frac{1}{4} =$ ____ (c) $\frac{1}{3} - \frac{1}{4} =$ ____

7

---

125. Complete: (a) $\frac{3}{10} \div \frac{2}{5} =$ ____

(b) $\frac{3}{10} + \frac{2}{5} =$ ____ (c) $\frac{3}{10} \times \frac{2}{5} =$ ____

a) $\frac{7}{12}$ b) $\frac{1}{12}$ c) $\frac{1}{12}$

---

126. Complete: (a) $\frac{5}{8} - \frac{1}{2} =$ ____

(b) $\frac{5}{8} \times \frac{1}{2} =$ ____ (c) $\frac{5}{8} \div \frac{1}{2} =$ ____

a) $\frac{3}{4}$ b) $\frac{7}{10}$ c) $\frac{3}{25}$

---

127. When the two fractions involved are identical, there is usually no problem with addition. However, when multiplying, there is a tendency to multiply the numerators <u>but to forget to multiply the denominators</u>. For example:

$$\frac{2}{3} + \frac{2}{3} = \frac{2+2}{3} = \frac{4}{3} = 1\frac{1}{3} \qquad\qquad \frac{2}{3} \times \frac{2}{3} = \frac{2 \times 2}{3 \times 3} = \underline{\quad}$$

a) $\frac{1}{8}$ b) $\frac{5}{16}$ c) $1\frac{1}{4}$

---

128. When the two fractions are identical, their <u>difference is "0"</u> and their <u>quotient is "1"</u>. For example:

$$\frac{2}{3} - \frac{2}{3} = \frac{2-2}{3} = \frac{0}{3} = 0 \qquad\qquad \frac{2}{3} \div \frac{2}{3} = \frac{2}{3} \times \frac{3}{2} = \underline{\quad}$$

$\frac{4}{9}$

1

129. In the next few frames, we will give some mixed practice with the four operations when the two fractions involved are identical.

Complete: (a) $\frac{2}{5} + \frac{2}{5} =$ _____ (b) $\frac{2}{5} \times \frac{2}{5} =$ _____ (c) $\frac{2}{5} - \frac{2}{5} =$ _____

130. Complete: (a) $\frac{1}{4} \div \frac{1}{4} =$ _____

(b) $\frac{1}{4} \times \frac{1}{4} =$ _____ (c) $\frac{1}{4} + \frac{1}{4} =$ _____

a) $\frac{4}{5}$ b) $\frac{4}{25}$ c) 0

131. Complete: (a) $\frac{7}{5} \times \frac{7}{5} =$ _____

(b) $\frac{7}{5} - \frac{7}{5} =$ _____ (c) $\frac{7}{5} \div \frac{7}{5} =$ _____

a) 1   b) $\frac{1}{16}$ c) $\frac{1}{2}$

132. We have added and multiplied $\frac{3}{4}$, $\frac{3}{4}$, and $\frac{3}{4}$ at the right.

$\frac{3}{4} + \frac{3}{4} + \frac{3}{4} = \frac{9}{4} = 2\frac{1}{4}$

Note: (1) To perform the addition, we simply added the three numerators.

$\frac{3}{4} \times \frac{3}{4} \times \frac{3}{4} = \frac{27}{64}$

a) $1\frac{24}{25}$ b) 0   c) 1

(2) To perform the multiplication, we multiplied both the three numerators and the three denominators. (Note: There is a tendency to forget to multiply the three denominators.)

Complete: (a) $\frac{1}{2} + \frac{1}{2} + \frac{1}{2} =$ _____ (b) $\frac{1}{2} \times \frac{1}{2} \times \frac{1}{2} =$ _____

a) $1\frac{1}{2}$ b) $\frac{1}{8}$

133. Complete: (a) $\frac{2}{3} \times \frac{2}{3} \times \frac{2}{3} =$ _____ (b) $\frac{2}{3} + \frac{2}{3} + \frac{2}{3} =$ _____

a) $\frac{8}{27}$ b) 2, from $\frac{6}{3}$

4-19   MIXED APPLIED PROBLEMS

This section contains some mixed applied problems. That is, each of the four basic operations involving fractions is needed to solve one or more problems.

134. A woman bought $\frac{7}{8}$ of a yard of red cloth and $\frac{5}{6}$ of a yard of blue cloth. Find the total amount of cloth bought.

$1\frac{17}{24}$ yard, from $\frac{7}{8} + \frac{5}{6}$

135. If a hiker is walking at the rate of 5 miles an hour, how far will he walk in $\frac{7}{10}$ of an hour?

$3\frac{1}{2}$ miles, from $\frac{7}{10} \times 5$

136. A man had $\frac{3}{4}$ of a gallon of gasoline. He used $\frac{5}{8}$ of a gallon to cut the grass. How much gasoline was left?

$\frac{1}{8}$ gallon, from $\frac{3}{4} - \frac{5}{8}$

137. If a girl's normal walking step is $\frac{3}{4}$ of a yard, how many steps must she take to walk 30 yards?

40 steps, from $30 \div \frac{3}{4}$

138. If a 1-inch length on a map represents 40 miles, find the number of miles represented by a $\frac{5}{16}$-inch length.

$12\frac{1}{2}$ miles, from $\frac{5}{16} \times 40$

139. A woman bought a $\frac{3}{4}$-pound steak that contained $\frac{1}{8}$ of a pound of fat. How much lean meat did it contain?

$\frac{5}{8}$ pound, from $\frac{3}{4} - \frac{1}{8}$

140. If a $\frac{1}{2}$-foot piece of wire is divided into 10 equal parts, how long is each part?

$\frac{1}{20}$ foot, from $\frac{1}{2} \div 10$

141. The amount of rainfall was $\frac{1}{8}$ of an inch during the morning and $\frac{1}{2}$ of an inch during the afternoon. Find the total amount of rainfall on that day.

$\frac{5}{8}$ inch, from $\frac{1}{8} + \frac{1}{2}$

142. A family has a yearly income of $15,000 and decides to save $\frac{1}{10}$ of it.

(a) How much will they save in one year?

(b) On the average, how much must they save each month?

a) $1,500, from $\frac{1}{10} \times \$15,000$ b) $125

143. A woman bought $\frac{5}{8}$ of a yard of nylon cloth and $\frac{9}{16}$ of a yard of rayon cloth.

(a) Did she buy more nylon or rayon?

(b) How much more of it did she buy?

a) nylon   b) $\frac{1}{16}$ yard, from $\frac{5}{8} - \frac{9}{16}$

144. A car travels $\frac{5}{6}$ of a mile in one minute.

(a) How far will it travel in 20 minutes?

(b) How far will it travel in $\frac{3}{4}$ of a minute?

a) $16\frac{2}{3}$ miles, from $20 \times \frac{5}{6}$

b) $\frac{5}{8}$ mile, from $\frac{3}{4} \times \frac{5}{6}$

145. In three months, Joe grew $\frac{3}{4}$ of an inch and Bill grew $\frac{13}{16}$ of an inch.

(a) Did Joe or Bill grow more?

(b) How much more did he grow?

a) Bill   b) $\frac{1}{16}$ inch, from $\frac{13}{16} - \frac{3}{4}$

SELF-TEST 10 (Frames 101-145)

Do the following multiplications.

1. $\frac{2}{3} \times \frac{6}{5} \times \frac{5}{4} =$
2. $\frac{1}{2}\left(\frac{4}{3}\right)\left(\frac{3}{10}\right) =$
3. $\left(\frac{7}{2}\right)\left(\frac{3}{4}\right)\left(\frac{2}{3}\right) =$

Do the following multiplications.

4. $\frac{1}{8} \times \frac{2}{5} \times 6 =$
5. $5 \times 4 \times \frac{7}{10} =$
6. $\left(\frac{3}{4}\right)(12)\left(\frac{5}{6}\right) =$

Do the following problems.

7. $\frac{5}{6} - \frac{1}{3} =$
8. $\frac{5}{6} \div \frac{1}{3} =$
9. $\frac{1}{2} + \frac{7}{8} =$
10. $\frac{1}{2} \times \frac{7}{8} =$

ANSWERS:

1. 1
2. $\frac{1}{5}$
3. $1\frac{3}{4}$
4. $\frac{3}{10}$
5. 14
6. $7\frac{1}{2}$
7. $\frac{1}{2}$
8. $2\frac{1}{2}$
9. $1\frac{3}{8}$
10. $\frac{7}{16}$

Unit 5 OPERATIONS WITH MIXED NUMBERS

In this unit, we will discuss additions, subtractions, multiplications, and divisions involving mixed numbers. Most of the principles used for the operations with mixed numbers are the same as those used for the operations with fractions.

5-1 ADDING MIXED NUMBERS WITH "LIKE" DENOMINATORS

In this section, we will discuss the procedure for adding mixed numbers with "like" denominators. Both the "improper fraction" method and the "whole number-fraction" method are discussed.

1. One method for adding mixed numbers involves converting each one to an improper fraction. We have used the "improper fraction" method to add $2\frac{1}{5}$ and $1\frac{3}{5}$ below. The steps are described.

    (1) Convert each mixed number to an improper fraction. $\quad 2\frac{1}{5} + 1\frac{3}{5} = \frac{11}{5} + \frac{8}{5}$

    (2) Add the improper fractions. $\quad 2\frac{1}{5} + 1\frac{3}{5} = \frac{11}{5} + \frac{8}{5} = \frac{19}{5}$

    (3) Convert the sum back to a mixed number. $\quad 2\frac{1}{5} + 1\frac{3}{5} = \frac{11}{5} + \frac{8}{5} = \frac{19}{5} = 3\frac{4}{5}$

    Following the steps above, complete the addition at the right. $\quad 2\frac{2}{7} + 1\frac{4}{7} = \frac{16}{7} + \frac{11}{7} = \underline{} = \underline{}$

2. When mixed numbers are added by the "improper fraction" method, the sum is always converted back to a mixed number. For example:

    $\frac{27}{7} = 3\frac{6}{7}$

    $4\frac{2}{5} + 3\frac{1}{5} = \frac{22}{5} + \frac{16}{5} = \frac{38}{5} = 7\frac{3}{5} \qquad 1\frac{2}{9} + 2\frac{2}{9} = \underline{} + \underline{} = \underline{} = \underline{}$

3. When converting a sum back to a mixed number, the fraction part of the mixed number is always reduced to lowest terms if possible. For example:

    $\frac{11}{9} + \frac{20}{9} = \frac{31}{9} = 3\frac{4}{9}$

    $2\frac{1}{4} + 3\frac{1}{4} = \frac{9}{4} + \frac{13}{4} = \frac{22}{4} = 5\frac{2}{4} = 5\frac{1}{2} \qquad 1\frac{3}{8} + 2\frac{3}{8} = \underline{} + \underline{} = \underline{} = \underline{}$

    $\frac{11}{8} + \frac{19}{8} = \frac{30}{8} = 3\frac{3}{4}$

4. Adding mixed numbers by the "improper fraction" method is tedious when either the whole numbers or denominators are large. For example:

$$11\frac{1}{3} + 12\frac{1}{3} = \frac{34}{3} + \frac{37}{3} = \frac{71}{3} = 23\frac{2}{3} \qquad 5\frac{3}{16} + 3\frac{5}{16} = \underline{\qquad} + \underline{\qquad} = \underline{\qquad}$$

$$\frac{83}{16} + \frac{53}{16} = \frac{136}{16} = 8\frac{1}{2}$$

5. Since adding mixed numbers by the "improper fraction" method is frequently tedious, we usually use a simpler method. The simpler method is described below.

Here are two additions performed by the "improper fraction" method in the last few frames.

$$2\frac{2}{7} + 1\frac{4}{7} = 3\frac{6}{7} \qquad\qquad 11\frac{1}{3} + 12\frac{1}{3} = 23\frac{2}{3}$$

We can find the sums in each case by simply adding the whole-number parts and adding the fraction parts of the original mixed numbers. That is:

$$2\frac{2}{7} + 1\frac{4}{7} = (2+1) + \left(\frac{2}{7} + \frac{4}{7}\right) \qquad 11\frac{1}{3} + 12\frac{1}{3} = (11+12) + \left(\frac{1}{3} + \frac{1}{3}\right)$$
$$= 3 + \frac{6}{7} = 3\frac{6}{7} \qquad\qquad = 23 + \frac{2}{3} = 23\frac{2}{3}$$

Did we obtain the same sum in each case by using the simpler method? _____

6. The simpler method for adding mixed numbers is called the "whole number - fraction" method. Use the "whole number - fraction" method to find each sum below.

Yes

(a) $8\frac{1}{5} + 6\frac{2}{5} = (8+6) + \left(\frac{1}{5} + \frac{2}{5}\right)$ (b) $1\frac{10}{13} + 6\frac{1}{13} = (1+6) + \left(\frac{10}{13} + \frac{1}{13}\right)$

$= \underline{\qquad} + \underline{\qquad} = \underline{\qquad}$ $= \underline{\qquad} + \underline{\qquad} = \underline{\qquad}$

7. When using the "whole number - fraction" method, we immediately get the sum in mixed-number form. Be sure to reduce the sum to lowest terms. For example:

a) $14 + \frac{3}{5} = 14\frac{3}{5}$ b) $7 + \frac{11}{13} = 7\frac{11}{13}$

$$7\frac{1}{6} + 8\frac{1}{6} = (7+8) + \left(\frac{1}{6} + \frac{1}{6}\right) \qquad 4\frac{5}{16} + 7\frac{7}{16} = (4+7) + \left(\frac{5}{16} + \frac{7}{16}\right)$$
$$= 15 + \frac{2}{6} = 15\frac{1}{3} \qquad\qquad = \underline{\qquad} + \underline{\qquad} = \underline{\qquad}$$

8. Use the "whole number - fraction" method for each addition below.

$11 + \frac{12}{16} = 11\frac{3}{4}$

(a) $3\frac{5}{12} + 4\frac{1}{12} =$ (b) $7\frac{1}{8} + 5\frac{1}{8} =$

a) $7\frac{1}{2}$ b) $12\frac{1}{4}$

5-2   ADDING MIXED NUMBERS WITH "UNLIKE" DENOMINATORS

In this section, we will discuss the procedure for adding mixed numbers with "unlike" denominators. Only the "whole number - fraction" method will be used.

9. The fraction parts of the mixed numbers at the left below have "unlike" denominators. We had to get common denominators before adding them. Using the same method, complete the other addition.

$$3\frac{1}{2} + 5\frac{1}{4} = (3+5) + \left(\frac{1}{2} + \frac{1}{4}\right) \qquad 7\frac{2}{3} + 2\frac{1}{6} = (7+2) + \left(\frac{2}{3} + \frac{1}{6}\right)$$
$$= 8 + \left(\frac{2}{4} + \frac{1}{4}\right) = 8\frac{3}{4} \qquad = 9 + \left(\underline{} + \frac{1}{6}\right) = \underline{}$$

10. Be sure to reduce the sum to lowest terms. For example:

$$9 + \left(\frac{4}{6} + \frac{1}{6}\right) = 9\frac{5}{6}$$

$$3\frac{1}{4} + 2\frac{1}{12} = (3+2) + \left(\frac{1}{4} + \frac{1}{12}\right) \qquad 7\frac{1}{6} + 1\frac{1}{2} = (7+1) + \left(\frac{1}{6} + \frac{1}{2}\right)$$
$$= 5 + \left(\frac{3}{12} + \frac{1}{12}\right) = 5\frac{1}{3} \qquad = 8 + \left(\frac{1}{6} + \underline{}\right) = \underline{}$$

11. The lowest common denominator was used to add the fractions at the left below. Use the same method to complete the other addition.

$$8 + \left(\frac{1}{6} + \frac{3}{6}\right) = 8\frac{2}{3}$$

$$5\frac{1}{6} + 2\frac{1}{4} = (5+2) + \left(\frac{1}{6} + \frac{1}{4}\right) \qquad 4\frac{2}{5} + 6\frac{1}{2} = (4+6) + \left(\frac{2}{5} + \frac{1}{2}\right)$$
$$= 7 + \left(\frac{2}{12} + \frac{3}{12}\right) = 7\frac{5}{12} \qquad = 10 + \left(\underline{} + \underline{}\right) = \underline{}$$

12. Complete: (a) $15\frac{3}{8} + 15\frac{1}{10} = $ _____

$$10 + \left(\frac{4}{10} + \frac{5}{10}\right) = 10\frac{9}{10}$$

(b) $13\frac{7}{24} + 18\frac{3}{8} = $ _____

a) $30\frac{19}{40}$ b) $31\frac{2}{3}$ (from $31\frac{16}{24}$)

5-3 ADDING A MIXED NUMBER TO A WHOLE NUMBER OR A FRACTION

In this section, we will briefly discuss the procedure for adding a mixed number to a whole number or to a proper fraction.

13. Each addition below involves a mixed number and a whole number. We have performed each by adding the whole number and the whole-number part of the mixed number.

$$3 + 2\frac{1}{4} = (3+2) + \frac{1}{4} = 5 + \frac{1}{4} = 5\frac{1}{4} \qquad 7\frac{2}{5} + 1 = (7+1) + \frac{2}{5} = 8 + \frac{2}{5} = 8\frac{2}{5}$$

Find each sum. (a) $4 + 3\frac{1}{3} = $ _____ (b) $5\frac{1}{2} + 5 = $ _____

14. Each addition below involves a mixed number and a fraction. We have performed each by adding the fraction and the fraction-part of the mixed number.

a) $7\frac{1}{3}$ b) $10\frac{1}{2}$

$$2\frac{1}{7} + \frac{4}{7} = 2 + \left(\frac{1}{7} + \frac{4}{7}\right) = 2 + \frac{5}{7} = 2\frac{5}{7} \qquad \frac{2}{5} + 8\frac{1}{5} = 8 + \left(\frac{2}{5} + \frac{1}{5}\right) = 8 + \frac{3}{5} = 8\frac{3}{5}$$

Find each sum. (a) $5\frac{1}{3} + \frac{1}{3} = $ _____ (b) $\frac{7}{9} + 6\frac{1}{9} = $ _____

15. In order to add the fraction and the fraction-part of the mixed number in each addition below, we have to get a common denominator first. For example:

$$1\frac{1}{2} + \frac{3}{8} = 1 + \left(\frac{1}{2} + \frac{3}{8}\right)$$
$$= 1 + \left(\frac{4}{8} + \frac{3}{8}\right) = 1\frac{7}{8}$$

$$\frac{2}{5} + 7\frac{1}{3} = 7 + \left(\frac{2}{5} + \frac{1}{3}\right)$$
$$= 7 + \left(\underline{} + \underline{}\right) = \underline{}$$

a) $5\frac{2}{3}$ b) $6\frac{8}{9}$

16. Be sure to reduce the sum to lowest terms. For example:

$$\frac{1}{6} + 9\frac{7}{10} = 9 + \left(\frac{1}{6} + \frac{7}{10}\right)$$
$$= 9 + \left(\frac{5}{30} + \frac{21}{30}\right) = 9\frac{13}{15}$$

$$1\frac{1}{6} + \frac{1}{3} = 1 + \left(\frac{1}{6} + \frac{1}{3}\right)$$
$$= 1 + \left(\underline{} + \underline{}\right) = \underline{}$$

$7 + \left(\frac{6}{15} + \frac{5}{15}\right) = 7\frac{11}{15}$

$1 + \left(\frac{1}{6} + \frac{2}{6}\right) = 1\frac{1}{2}$

17. Complete: (a) $27\frac{7}{16} + 42 = $ \underline{}   (b) $\frac{7}{24} + 56\frac{3}{8} = $ \underline{}

a) $69\frac{7}{16}$ b) $56\frac{2}{3}$

5-4   MIXED-NUMBER SUMS THAT REQUIRE A REGROUPING

In additions involving mixed numbers, we can obtain a sum in which the fraction part is an improper fraction. Sums of this type are not in the "correct" form. In this section, we will discuss the regrouping procedure needed to convert them to the "correct" form.

18. In the "correct" form for mixed numbers, the fraction part is a <u>proper</u> fraction. That is:

$4\frac{1}{3}$, $2\frac{5}{6}$, and $5\frac{7}{8}$ <u>are</u> in the correct form.

$2\frac{5}{3}$, $9\frac{7}{6}$, and $1\frac{11}{8}$ <u>are not</u> in the correct form.

Which of these mixed numbers are in the correct form? \underline{}   (a) $10\frac{3}{4}$ (b) $2\frac{7}{5}$ (c) $6\frac{13}{12}$ (d) $1\frac{9}{10}$

19. We have used the "whole number - fraction" method to obtain each sum below. Notice that the fraction part of each sum is an <u>improper</u> fraction.

$$4\frac{2}{3} + 1\frac{2}{3} = 5\frac{4}{3}$$
$$3\frac{4}{7} + 5\frac{6}{7} = 8\frac{10}{7}$$

Only (a) and (d)

Since $\frac{4}{3}$ and $\frac{10}{7}$ are improper fractions, the sums above <u>are not</u> in the correct form. To put them in the correct form, a regrouping procedure is needed. The regrouping procedure for $5\frac{4}{3}$ is shown below.

$$5\frac{4}{3} = 5 + \frac{4}{3} = 5 + 1\frac{1}{3} = 6\frac{1}{3}$$

Note: First we converted $\frac{4}{3}$ to $1\frac{1}{3}$. Then we added 5 to $1\frac{1}{3}$.

Using the same procedure, convert $8\frac{10}{7}$ to the correct form.

$$8\frac{10}{7} = 8 + \frac{10}{7} = 8 + \underline{} = \underline{}$$

$8 + 1\frac{3}{7} = 9\frac{3}{7}$

20. Convert each of these to the correct form.   (a) $2\frac{5}{4} = 2 + 1\frac{1}{4} = $ _____   (b) $1\frac{11}{9} = 1 + 1\frac{2}{9} = $ _____

21. Convert each of the following to the correct form. | a) $3\frac{1}{4}$ b) $2\frac{2}{9}$

(a) $5\frac{9}{5} = 5 + $ _____ $= $ _____   (b) $10\frac{5}{3} = 10 + $ _____ $= $ _____

22. Convert each of the following to the correct form. | a) $5 + 1\frac{4}{5} = 6\frac{4}{5}$ b) $10 + 1\frac{2}{3} = 11\frac{2}{3}$

(a) $7\frac{11}{8} = $ _____   (b) $1\frac{7}{4} = $ _____   (c) $34\frac{17}{16} = $ _____

23. Perform each addition below and convert each sum to the correct form. | a) $8\frac{3}{8}$ b) $2\frac{3}{4}$ c) $35\frac{1}{16}$

(a) $3\frac{4}{5} + 5\frac{2}{5} = $ _____   (b) $\frac{8}{9} + 1\frac{8}{9} = $ _____

24. When converting a mixed-number sum to the correct form, be sure to reduce the fraction part to lowest terms. For example: | a) $9\frac{1}{5}$ (from $8\frac{6}{5}$)

$4\frac{10}{6} = 5\frac{4}{6} = 5\frac{2}{3}$ $7\frac{6}{4} = 8\frac{2}{4} = 8\frac{1}{2}$ | b) $2\frac{7}{9}$ (from $1\frac{16}{9}$)

Convert each of the following to the correct form.

(a) $5\frac{12}{9} = $ _____   (b) $9\frac{14}{8} = $ _____   (c) $10\frac{14}{10} = $ _____

25. Complete each addition. Write each sum in the correct form. | a) $6\frac{1}{3}$ (from $6\frac{3}{9}$)   b) $10\frac{3}{4}$ (from $10\frac{6}{8}$)   c) $11\frac{2}{5}$ (from $11\frac{4}{10}$)

(a) $1\frac{7}{12} + 2\frac{7}{12} = $ _____   (b) $10\frac{2}{3} + 20\frac{5}{6} = $ _____

26. Write each sum in the correct form. | a) $4\frac{1}{6}$ b) $31\frac{1}{2}$

(a) $1\frac{3}{5} + \frac{5}{6} = $ _____   (b) $\frac{7}{8} + 57\frac{5}{12} = $ _____

27. In each sum below, the fraction part is an improper fraction that equals "1". | a) $2\frac{13}{30}$ b) $58\frac{7}{24}$

$5\frac{3}{4} + 2\frac{1}{4} = 7\frac{4}{4}$ $3\frac{5}{8} + 1\frac{3}{8} = 4\frac{8}{8}$

We have converted each sum above to the correct form below. Notice that the correct form for each is simply a whole number.

$7\frac{4}{4} = 7 + \frac{4}{4} = 7 + 1 = 8$ $4\frac{8}{8} = 4 + \frac{8}{8} = 4 + 1 = 5$

Convert each of these to the correct form.   (a) $1\frac{7}{7} = $ _____   (b) $3\frac{5}{5} = $ _____   (c) $42\frac{16}{16} = $ _____

| a) 2   b) 4   c) 43

28. Write each sum below in the correct form.

(a) $3\frac{5}{9} + 2\frac{4}{9} =$ _____

(b) $\frac{17}{32} + 7\frac{15}{32} =$ _____

a) 6   b) 8

5-5 ADDING THREE MIXED NUMBERS

In this section, we will discuss additions involving three mixed numbers. Mixed numbers with both "like" and "unlike" denominators are included.

29. To add three mixed numbers, we can also use the "whole number – fraction" method. That is, we simply add the three whole numbers and add the three fractions. For example:

$$2\frac{1}{7} + 1\frac{3}{7} + 4\frac{2}{7} = (2 + 1 + 4) + \left(\frac{1}{7} + \frac{3}{7} + \frac{2}{7}\right) = 7 + \frac{6}{7} = 7\frac{6}{7}$$

Use the "whole-number – fraction" method to find the sum at the right.

$$4\frac{1}{5} + 3\frac{2}{5} + 1\frac{1}{5} = (4 + 3 + 1) + \left(\frac{1}{5} + \frac{2}{5} + \frac{1}{5}\right) = \underline{} + \underline{} = \underline{}$$

30. When adding three mixed numbers, the fraction part of the sum can be an improper fraction. The sum must then be converted to the correct form. For example:

$8 + \frac{4}{5} = 8\frac{4}{5}$

$$2\frac{1}{6} + 3\frac{5}{6} + 4\frac{1}{6} = (2 + 3 + 4) + \left(\frac{1}{6} + \frac{5}{6} + \frac{1}{6}\right) = 9 + \frac{7}{6} = 9\frac{7}{6} = 10\frac{1}{6}$$

Convert the sum at the right to the correct form.

$$1\frac{2}{3} + 5\frac{1}{3} + 6\frac{2}{3} = (1 + 5 + 6) + \left(\frac{2}{3} + \frac{1}{3} + \frac{2}{3}\right) = \underline{} + \underline{} = \underline{}$$

31. When adding three mixed numbers, the fraction part of the sum can be an improper fraction that equals the number "1". The sum must then be converted to a whole number. For example:

$12 + \frac{5}{3} = 13\frac{2}{3}$

$$2\frac{1}{7} + 1\frac{4}{7} + 3\frac{2}{7} = (2 + 1 + 3) + \left(\frac{1}{7} + \frac{4}{7} + \frac{2}{7}\right) = 6 + \frac{7}{7} = 6\frac{7}{7} = 7$$

Convert the sum at the right to a whole number.

$$2\frac{6}{11} + 3\frac{2}{11} + 8\frac{3}{11} = (2 + 3 + 8) + \left(\frac{6}{11} + \frac{2}{11} + \frac{3}{11}\right) = \underline{} + \underline{} = \underline{}$$

$13 + \frac{11}{11} = 14$

101

32. In the addition below, the fraction part of the sum is an improper fraction that is larger than the whole number "2". We have converted it to the correct form.

$$2\frac{7}{8} + 3\frac{5}{8} + 4\frac{7}{8} = 9 + \frac{19}{8} = 9 + 2\frac{3}{8} = 11\frac{3}{8}$$

Convert each of these to the correct form. (a) $8\frac{15}{7} = $ _____ (b) $4\frac{13}{5} = $ _____

33. Convert the sum at the right to the correct form. $2\frac{9}{10} + 5\frac{7}{10} + 6\frac{7}{10} = $ _____

a) $10\frac{1}{7}$ b) $6\frac{3}{5}$

34. In the addition below, the fraction part of the sum is an improper fraction that equals the whole number "2". We have converted the sum to a whole number.

$15\frac{3}{10}$ (from $13\frac{23}{10}$)

$$1\frac{3}{5} + 3\frac{4}{5} + 2\frac{3}{5} = 6 + \frac{10}{5} = 6 + 2 = 8$$

Convert each sum to a whole number. (a) $5\frac{4}{7} + 1\frac{5}{7} + 3\frac{5}{7} = $ _____ (b) $4\frac{8}{9} + 8\frac{2}{9} + 1\frac{8}{9} = $ _____

35. When adding three mixed numbers, remember that the sum must be reduced to lowest terms if possible. For example:

a) 11 (from $9\frac{14}{7}$)   b) 15 (from $13\frac{18}{9}$)

$$7\frac{1}{6} + 8\frac{1}{6} + 5\frac{1}{6} = 20\frac{3}{6} = 20\frac{1}{2} \qquad 2\frac{8}{9} + 4\frac{5}{9} + 3\frac{8}{9} = $$ _____

36. The additions below contain mixed numbers together with whole numbers and proper fractions. The "whole number - fraction" method can still be used. That is:

$11\frac{1}{3}$ (from $9\frac{21}{9}$)

$$3\frac{1}{5} + 2 + 5\frac{3}{5} = (3 + 2 + 5) + \left(\frac{1}{5} + \frac{3}{5}\right) = 10\frac{4}{5}$$

$$1\frac{4}{7} + 4\frac{2}{7} + \frac{5}{7} = (1 + 4) + \left(\frac{4}{7} + \frac{2}{7} + \frac{5}{7}\right) = 5\frac{11}{7} = 6\frac{4}{7}$$

$$10 + 3\frac{7}{8} + \frac{1}{8} = (10 + 3) + \left(\frac{7}{8} + \frac{1}{8}\right) = 13\frac{8}{8} = 14$$

Perform each addition below. Report each sum in the correct form.

(a) $4\frac{8}{9} + 3\frac{1}{9} + 1 = $ _____   (b) $\frac{5}{6} + 7 + 2\frac{5}{6} = $ _____   (c) $1\frac{5}{7} + \frac{3}{7} + 5\frac{6}{7} = $ _____

37. To add the fractions at the left below, we used 8 as the common denominator, since 8 is a multiple of both 4 and 2. Complete the other addition.

a) 9 (from $8\frac{9}{9}$)   b) $10\frac{2}{3}$ (from $9\frac{10}{6}$)   c) 8 (from $6\frac{14}{7}$)

$$2\frac{1}{8} + 1\frac{1}{4} + 3\frac{1}{2} = 6 + \left(\frac{1}{8} + \frac{2}{8} + \frac{4}{8}\right) = 6\frac{7}{8} \qquad 2\frac{1}{6} + 4\frac{1}{2} + 6\frac{2}{3} = $$ _____

$13\frac{1}{3}$ (from $12\frac{8}{6}$)

38. For the addition below, we used 24 as the common denominator since 24 is the smallest multiple of 8 that is also a multiple of both 4 and 6. Complete the other addition.

$5\frac{1}{4} + 2\frac{3}{8} + \frac{1}{6} = 7 + \left(\frac{6}{24} + \frac{9}{24} + \frac{4}{24}\right) = 7\frac{19}{24}$ $1\frac{3}{4} + \frac{5}{6} + 2\frac{3}{4} =$ _____ | $5\frac{1}{3}$

5-6 APPLIED PROBLEMS

This section contains some verbal or applied problems. All of them can be solved by an addition involving mixed numbers.

39. $19\frac{1}{2}$ yards of carpeting are needed for one room. $14\frac{1}{2}$ yards of carpeting are needed for a second room. Find the total amount of carpeting needed.

| 34 yards

40. In one month, a housewife bought gasoline for her car twice. If she bought $14\frac{1}{2}$ gallons and $17\frac{9}{10}$ gallons, find the total amount of gasoline bought.

| $32\frac{2}{5}$ gallons

41. On a part-time job, a student worked $3\frac{1}{2}$ hours on Thursday, $2\frac{3}{4}$ hours on Friday, and 6 hours on Saturday. Find her total number of hours worked in the three days.

| $12\frac{1}{4}$ hours

42. A fruit stand sold $8\frac{1}{2}$ bushels of apples on Friday, $12\frac{3}{4}$ bushels on Saturday, and $23\frac{1}{4}$ bushels on Sunday. Find the total number of bushels sold.

| $44\frac{1}{2}$ bushels

SELF-TEST 11 (Frames 1-42)

Do the following additions.

1. $5\frac{7}{10} + 8\frac{9}{10} =$

2. $4\frac{3}{4} + 3\frac{1}{6} =$

3. $1\frac{5}{6} + 7 =$

4. $\frac{11}{16} + 2\frac{3}{8} =$

5. $3\frac{9}{10} + 1\frac{4}{15} =$

6. $2\frac{7}{8} + 1\frac{1}{2} + 4\frac{5}{8} =$

ANSWERS: 1. $14\frac{3}{5}$ 2. $7\frac{11}{12}$ 3. $8\frac{5}{6}$ 4. $3\frac{1}{16}$ 5. $5\frac{1}{6}$ 6. 9

5-7 SUBTRACTING MIXED NUMBERS

In this section, we will discuss the procedure for subtracting mixed numbers with both "like" and "unlike" denominators. Though the "improper fraction" method is shown, the "whole number - fraction" method is emphasized.

43. We used the "improper fraction" method to perform the subtraction at the left. That is, we began by converting each mixed number to an improper fraction. Use the same method to perform the subtraction at the right. Be sure to convert the difference back to a mixed number.

$$5\frac{3}{5} - 2\frac{1}{5} = \frac{28}{5} - \frac{11}{5} = \frac{17}{5} = 3\frac{2}{5} \qquad 3\frac{2}{3} - 1\frac{1}{3} = \underline{} - \underline{} = \underline{} = \underline{}$$

44. Just as adding mixed numbers by the "improper fraction" method is tedious when the whole numbers or denominators are large, subtracting mixed numbers by that method is tedious with large numbers. For example:

$$\frac{11}{3} - \frac{4}{3} = \frac{7}{3} = 2\frac{1}{3}$$

$$24\frac{2}{3} - 10\frac{1}{3} = \frac{74}{3} - \frac{31}{3} = \frac{43}{3} = 14\frac{1}{3} \qquad 4\frac{15}{16} - 3\frac{7}{16} = \frac{79}{16} - \frac{55}{16} = \underline{} = \underline{}$$

45. Here are the two subtractions performed by the "improper fraction" method in the last frame.

$$\frac{24}{16} = 1\frac{1}{2}$$

$$24\frac{2}{3} - 10\frac{1}{3} = 14\frac{1}{3} \qquad 4\frac{15}{16} - 3\frac{7}{16} = 1\frac{8}{16} = 1\frac{1}{2}$$

We can find the differences more easily by using the "whole number - fraction" method. That is:

$$24\frac{2}{3} - 10\frac{1}{3} = (24 - 10) + \left(\frac{2}{3} - \frac{1}{3}\right) = 14 + \frac{1}{3} = 14\frac{1}{3}$$

$$4\frac{15}{16} - 3\frac{7}{16} = (4 - 3) + \left(\frac{15}{16} - \frac{7}{16}\right) = 1 + \frac{8}{16} = 1\frac{8}{16} = 1\frac{1}{2}$$

Note: (1) We simply subtracted the two whole numbers and then subtracted the two fractions.

(2) We <u>added</u> the difference between the two whole numbers and the difference between the two <u>fractions</u>.

Use the "whole number - fraction" method to perform this subtraction.

$$9\frac{5}{7} - 4\frac{1}{7} = (9 - 4) + \left(\frac{5}{7} - \frac{1}{7}\right) = \underline{} + \underline{} = \underline{}$$

46. When performing subtractions, be sure to reduce the difference to lowest terms. For example:

$$5 + \frac{4}{7} = 5\frac{4}{7}$$

$$7\frac{7}{8} - 5\frac{5}{8} = (7 - 5) + \left(\frac{7}{8} - \frac{5}{8}\right) = 2\frac{2}{8} = 2\frac{1}{4} \qquad 35\frac{5}{6} - 25\frac{1}{6} = (35 - 25) + \left(\frac{5}{6} - \frac{1}{6}\right) = \underline{}$$

47. Use the "whole number - fraction" method to find each difference below.

$$10\frac{2}{3} \text{ (from } 10\frac{4}{6}\text{)}$$

(a) $100\frac{8}{9} - 50\frac{1}{9} = $

(b) $7\frac{31}{32} - 4\frac{7}{32} = $

a) $50\frac{7}{9}$ b) $3\frac{3}{4}$ (from $3\frac{24}{32}$)

104

48. When the mixed numbers have "unlike" denominators, we have to get a common denominator in order to subtract their fraction parts. For example:

$$4\frac{5}{8} - 2\frac{1}{4} = (4-2) + \left(\frac{5}{8} - \frac{1}{4}\right) \qquad 5\frac{2}{3} - 3\frac{1}{4} = (5-3) + \left(\frac{2}{3} - \frac{1}{4}\right)$$

$$= 2 + \left(\frac{5}{8} - \frac{2}{8}\right) = 2\frac{3}{8} \qquad = 2 + \left(\frac{8}{12} - \frac{3}{12}\right) = 2\frac{5}{12}$$

Following the examples above, complete each subtraction below.

(a) $5\frac{3}{5} - 1\frac{3}{10} = (5-1) + \left(\frac{3}{5} - \frac{3}{10}\right)$

$= 4 + \left(\underline{} - \frac{3}{10}\right) = \underline{}$

(b) $7\frac{3}{5} - 2\frac{1}{2} = (7-2) + \left(\frac{3}{5} - \frac{1}{2}\right)$

$= 5 + \left(\underline{} - \underline{}\right) = \underline{}$

49. Use the "whole number - fraction" method for each subtraction below. Be sure that each difference is reported in lowest terms.

(a) $5\frac{5}{6} - 3\frac{1}{2} = $ _____

(b) $25\frac{3}{4} - 15\frac{1}{10} = $ _____

a) $4 + \left(\frac{6}{10} - \frac{3}{10}\right) = 4\frac{3}{10}$

b) $5 + \left(\frac{6}{10} - \frac{5}{10}\right) = 5\frac{1}{10}$

50. When we subtract the whole numbers, we can get "0". In such cases, the difference is simply a proper fraction. For example:

$$9\frac{3}{5} - 9\frac{2}{5} = (9-9) + \left(\frac{3}{5} - \frac{2}{5}\right) = 0 + \frac{1}{5} = \frac{1}{5} \qquad 10\frac{15}{16} - 10\frac{11}{16} = $$ _____

a) $2\frac{1}{3}$ (from $2\frac{2}{6}$) b) $10\frac{13}{20}$

51. When we subtract the fractions, we can also get "0". In such cases, the difference is simply a whole number. For example:

$$10\frac{3}{5} - 4\frac{3}{5} = (10-4) + \left(\frac{3}{5} - \frac{3}{5}\right) = 6 + \frac{0}{5} = 6 \qquad 25\frac{7}{8} - 10\frac{7}{8} = $$ _____

$\frac{1}{4}$ (from $\frac{4}{16}$)

52. When subtracting a mixed number from itself, we get "0" for both the whole-number subtraction and the fraction subtraction. Therefore, the difference is "0". That is:

$$7\frac{3}{5} - 7\frac{3}{5} = 0 \qquad 10\frac{1}{4} - 10\frac{1}{4} = 0 \qquad 33\frac{2}{3} - 33\frac{2}{3} = $$ _____

15 (from $15 + \frac{0}{8}$)

0

5-8 SUBTRACTING MIXED NUMBERS WHEN "BORROWING" IS NEEDED

When the fraction part of the second number is larger than the fraction part of the first mixed number, "borrowing" is needed in the subtraction process. We will discuss that type of "borrowing" in this section.

53. In the subtraction below, $\frac{3}{7}$ is larger than $\frac{1}{7}$. Therefore, we cannot use the "whole number – fraction" method with the mixed numbers as they stand since $\frac{1}{7} - \frac{3}{7}$ cannot be done in arithmetic.

$$5\frac{1}{7} - 3\frac{3}{7} = (5-3) + \left(\frac{1}{7} - \frac{3}{7}\right)$$

In order to perform the subtraction, we must regroup $5\frac{1}{7}$ to an equivalent form in which the numerator of the fraction is larger than 3. We have done so below.

$$5\frac{1}{7} = 4 + 1 + \frac{1}{7} = 4 + \frac{7}{7} + \frac{1}{7} = 4\frac{8}{7}$$

Note: We "borrowed 1" from the 5 leaving 4, converted the "1" to $\frac{7}{7}$, and then added the $\frac{7}{7}$ to $\frac{1}{7}$ to get $\frac{8}{7}$.

Having regrouped the $5\frac{1}{7}$ to $4\frac{8}{7}$, we can complete the subtraction. That is:

$$5\frac{1}{7} - 3\frac{3}{7} = 4\frac{8}{7} - 3\frac{3}{7} = \underline{\qquad}$$

54. We have used the "borrowing" procedure to regroup each mixed number below. | $1\frac{5}{7}$

$$7\frac{2}{5} = 6 + \frac{5}{5} + \frac{2}{5} = 6\frac{7}{5} \qquad\qquad 2\frac{7}{8} = 1 + \frac{8}{8} + \frac{7}{8} = 1\frac{15}{8}$$

Use the same procedure to regroup each of the following.

(a) $4\frac{1}{2} = 3 + \underline{\qquad} + \frac{1}{2} = \underline{\qquad}$ (b) $9\frac{5}{6} = 8 + \underline{\qquad} + \frac{5}{6} = \underline{\qquad}$

55. By mentally using the "borrowing 1" procedure, write the correct numerator in each box below.

(a) $5\frac{2}{9} = 4\frac{\Box}{9}$ (c) $7\frac{7}{10} = 6\frac{\Box}{10}$

(b) $3\frac{1}{5} = 2\frac{\Box}{5}$ (d) $11\frac{3}{4} = 10\frac{\Box}{4}$

a) $3 + \frac{2}{2} + \frac{1}{2} = 3\frac{3}{2}$

b) $8 + \frac{6}{6} + \frac{5}{6} = 8\frac{11}{6}$

56. Using the "borrowing 1" procedure, complete each subtraction below.

(a) $3\frac{2}{5} - 1\frac{4}{5} = \underline{\qquad} - 1\frac{4}{5} = \underline{\qquad}$ (b) $8\frac{1}{9} - 5\frac{2}{9} = \underline{\qquad} - 5\frac{2}{9} = \underline{\qquad}$

a) $4\frac{11}{9}$ b) $2\frac{6}{5}$ c) $6\frac{17}{10}$ d) $10\frac{7}{4}$

57. Using the same procedure, complete each subtraction below.

(a) $10\frac{1}{3} - 3\frac{2}{3} = \underline{\qquad} - 3\frac{2}{3} = \underline{\qquad}$ (b) $7\frac{5}{11} - 1\frac{9}{11} = \underline{\qquad} - 1\frac{9}{11} = \underline{\qquad}$

a) $2\frac{7}{5} - 1\frac{4}{5} = 1\frac{3}{5}$ b) $7\frac{10}{9} - 5\frac{2}{9} = 2\frac{8}{9}$

a) $9\frac{4}{3} - 3\frac{2}{3} = 6\frac{2}{3}$ b) $6\frac{16}{11} - 1\frac{9}{11} = 5\frac{7}{11}$

106

58. Complete each subtraction below. Reduce each difference to lowest terms.

   (a) $8\frac{1}{4} - 3\frac{3}{4} = $ _____   (b) $10\frac{11}{16} - 8\frac{15}{16} = $ _____

59. When the mixed numbers have "unlike" denominators, we sometimes have to use the "borrowing 1" procedure after we get common denominators. For example:

   $5\frac{1}{3} - 1\frac{5}{6} = 5\frac{2}{6} - 1\frac{5}{6} = 4\frac{8}{6} - 1\frac{5}{6} = 3\frac{3}{6} = 3\frac{1}{2}$

   a) $4\frac{1}{2}$, from $7\frac{5}{4} - 3\frac{3}{4}$

   b) $1\frac{3}{4}$, from $9\frac{27}{16} - 8\frac{15}{16}$

   Note: (1) We began by converting $5\frac{1}{3}$ to $5\frac{2}{6}$ to get a common denominator.

   (2) Then we regrouped $5\frac{2}{6}$ to $4\frac{8}{6}$.

   Following the steps above, complete the following. $7\frac{3}{8} - 3\frac{1}{2} = 7\frac{3}{8} - 3\frac{4}{8} = $ _____ $- 3\frac{4}{8} = $ _____

60. To complete the subtraction below, we got common denominators before regrouping $4\frac{5}{15}$ to $3\frac{20}{15}$.

   $4\frac{1}{3} - 2\frac{2}{5} = 4\frac{5}{15} - 2\frac{6}{15} = 3\frac{20}{15} - 2\frac{6}{15} = 1\frac{14}{15}$

   $6\frac{11}{8} - 3\frac{4}{8} = 3\frac{7}{8}$

   Following the steps above, complete the following subtraction. $8\frac{1}{6} - 3\frac{7}{8} = 8\frac{4}{24} - 3\frac{21}{24} = $ _____ $- 3\frac{21}{24} = $ _____

61. Complete each of the following subtractions. Be sure to obtain common denominators before "borrowing".

   $7\frac{28}{24} - 3\frac{21}{24} = 4\frac{7}{24}$

   (a) $8\frac{1}{10} - 6\frac{1}{8} = $ _____   (b) $5\frac{1}{2} - 1\frac{2}{3} = $ _____

62. When we borrow "1" from the 7 in the subtraction at the left below, we leave 6. Since $6 - 6 = 0$, the whole-number part of the difference is "0". Therefore, the difference is simply a proper fraction. Complete the other subtraction.

   a) $1\frac{39}{40}$ b) $3\frac{5}{6}$

   $7\frac{1}{5} - 6\frac{4}{5} = 6\frac{6}{5} - 6\frac{4}{5} = 0 + \frac{2}{5} = \frac{2}{5}$ $5\frac{1}{8} - 4\frac{1}{4} = $ _____

   $\frac{7}{8}$

5-9   SUBTRACTING WHOLE NUMBERS AND FRACTIONS FROM MIXED NUMBERS

In this section, we will discuss the procedure for subtracting whole numbers and proper fractions from mixed numbers.

63. To subtract a whole number from a mixed number, we simply subtract the whole numbers. For example:

   $7\frac{1}{2} - 5 = (7 - 5) + \frac{1}{2} = 2\frac{1}{2}$ $35\frac{9}{16} - 25 = $ _____

   $10\frac{9}{16}$

64. When the whole number is the same as the whole-number part of the mixed number, the difference is simply a proper fraction. For example:

$8\frac{3}{4} - 8 = (8 - 8) + \frac{3}{4} = 0 + \frac{3}{4} = \frac{3}{4}$ $\qquad 1\frac{7}{8} - 1 = $ _____

65. To subtract a proper fraction from a mixed number, we simply subtract the fractions. For example:

$4\frac{5}{7} - \frac{2}{7} = 4 + \left(\frac{5}{7} - \frac{2}{7}\right) = 4\frac{3}{7}$ $\qquad 7\frac{3}{5} - \frac{2}{5} = $ _____

$\boxed{\frac{7}{8}}$

66. When the fraction is the same as the fraction part of the mixed number, the difference is simply a whole number. For example:

$3\frac{5}{6} - \frac{5}{6} = 3 + \left(\frac{5}{6} - \frac{5}{6}\right) = 3 + \frac{0}{6} = 3$ $\qquad 1\frac{1}{2} - \frac{1}{2} = $ _____

$\boxed{7\frac{1}{5}}$

67. Remember to reduce each difference to lowest terms. For example:

$3\frac{5}{6} - \frac{1}{6} = 3\frac{4}{6} = 3\frac{2}{3}$ $\qquad 10\frac{15}{16} - \frac{7}{16} = $ _____

$\boxed{1}$

68. To perform each subtraction below, we have to get a common denominator first. That is:

$5\frac{3}{4} - \frac{1}{2} = 5\frac{3}{4} - \frac{2}{4} = 5\frac{1}{4}$ $\qquad 7\frac{2}{3} - \frac{1}{4} = 7\frac{8}{12} - \frac{3}{12} = $ _____

$\boxed{10\frac{1}{2}}$

69. To perform each subtraction below, we have to "borrow 1" from the whole number. That is:

$3\frac{1}{7} - \frac{4}{7} = 2\frac{8}{7} - \frac{4}{7} = 2\frac{4}{7}$ $\qquad 4\frac{1}{8} - \frac{7}{8} = $ _____ $-$ _____ $=$ _____

$\boxed{7\frac{5}{12}}$

70. After getting a common denominator in each subtraction below, we have to "borrow 1" from the whole number. That is:

$5\frac{1}{3} - \frac{5}{6} = 5\frac{2}{6} - \frac{5}{6} = 4\frac{8}{6} - \frac{5}{6} = 4\frac{3}{6} = 4\frac{1}{2}$ $\qquad 8\frac{1}{2} - \frac{5}{7} = 8\frac{7}{14} - \frac{10}{14} = $ _____ $-$ _____ $=$ _____

$\boxed{3\frac{9}{8} - \frac{7}{8} = 3\frac{1}{4}}$

71. When we "borrow 1" from "1", we leave "0". Therefore, the difference is simply a proper fraction. For example:

$1\frac{1}{4} - \frac{1}{2} = 1\frac{1}{4} - \frac{2}{4} = \frac{5}{4} - \frac{2}{4} = \frac{3}{4}$ $\qquad 1\frac{1}{3} - \frac{3}{5} = $ _____

$\boxed{7\frac{21}{14} - \frac{10}{14} = 7\frac{11}{14}}$

72. Complete each of the following subtractions.

(a) $10\frac{2}{3} - 7 = $ _____

(c) $5\frac{5}{8} - 5 = $ _____

(b) $3\frac{1}{4} - \frac{3}{8} = $ _____

(d) $1\frac{1}{2} - \frac{3}{5} = $ _____

$\boxed{\frac{11}{15}}$

$\boxed{\text{a) } 3\frac{2}{3} \quad \text{c) } \frac{5}{8} \\ \text{b) } 2\frac{7}{8} \quad \text{d) } \frac{9}{10}}$

107

5-10 SUBTRACTING MIXED NUMBERS AND FRACTIONS FROM WHOLE NUMBERS

In this section, we will discuss the procedure for subtracting mixed numbers and proper fractions from whole numbers.

73. We cannot perform the subtraction $5 - 3\frac{6}{7}$ as it stands since "5" has no fraction part. In order to perform it, we must regroup 5 to an equivalent form containing a fraction part. We have done so below.

$$5 = 4 + 1 = 4 + \frac{7}{7} = 4\frac{7}{7}$$

Note: We "borrowed 1" from the 5 leaving 4, and then converted the "1" to $\frac{7}{7}$.

Having regrouped the 5 to $4\frac{7}{7}$, we can complete the subtraction. That is:

$$5 - 3\frac{6}{7} = 4\frac{7}{7} - 3\frac{6}{7} = \underline{\hspace{1cm}}$$

74. By mentally using the "borrowing 1" procedure, write the correct numerator in each box below.

(a) $3 = 2\frac{\Box}{8}$ (b) $7 = 6\frac{\Box}{3}$ (c) $9 = 8\frac{\Box}{10}$

$1\frac{1}{7}$

75. By using the "borrowing 1" procedure, we have completed the subtraction at the left below. Complete the other subtraction.

$$4 - 1\frac{1}{2} = 3\frac{2}{2} - 1\frac{1}{2} = 2\frac{1}{2} \qquad 7 - 2\frac{5}{8} = \underline{\hspace{1cm}}$$

a) $2\frac{8}{8}$ b) $6\frac{3}{3}$ c) $8\frac{10}{10}$

76. When we "borrowed 1" from the 9 in the subtraction at the left below, we left 8. Therefore, the difference is simply a proper fraction. Complete the other subtraction.

$$9 - 8\frac{2}{3} = 8\frac{3}{3} - 8\frac{2}{3} = \frac{1}{3} \qquad 20 - 19\frac{7}{9} = \underline{\hspace{1cm}}$$

$4\frac{3}{8}$

77. We also have to use the "borrowing 1" procedure to subtract a proper fraction from a whole number. For example:

$$3 - \frac{1}{5} = 2\frac{5}{5} - \frac{1}{5} = 2\frac{4}{5} \qquad 5 - \frac{3}{8} = \underline{\hspace{1cm}}$$

$\frac{2}{9}$

78. To perform the subtraction at the left below, we simply converted the "1" to $\frac{6}{6}$. Complete the other subtractions.

$$1 - \frac{5}{6} = \frac{6}{6} - \frac{5}{6} = \frac{1}{6} \qquad (a) \ 1 - \frac{2}{7} = \underline{\hspace{1cm}} \qquad (b) \ 1 - \frac{1}{8} = \underline{\hspace{1cm}}$$

$4\frac{5}{8}$

79. Complete each of the following subtractions.

(a) $4 - \frac{8}{9} =$

(b) $6 - 5\frac{3}{4} =$

(c) $10 - 2\frac{1}{3} =$

(d) $1 - \frac{3}{10} =$

a) $\frac{5}{7}$ b) $\frac{7}{8}$

a) $3\frac{1}{9}$ c) $7\frac{2}{3}$

b) $\frac{1}{4}$ d) $\frac{7}{10}$

5-11 COMPARING THE SIZE OF MIXED NUMBERS

In this section, we will briefly discuss the procedure for comparing the size of mixed numbers. We will show that common denominators are sometimes needed in order to make these comparisons.

80. If two mixed numbers have different whole-number parts, <u>the one with the larger whole number is larger</u>. For example:

    $5\frac{1}{4}$ is larger than $3\frac{2}{7}$, since 5 is larger than 3

    Draw a circle around the <u>larger</u> mixed number in each pair.   (a) $3\frac{3}{4}$ and $2\frac{1}{4}$ (b) $34\frac{6}{7}$ and $35\frac{1}{7}$

81. If two mixed numbers have the same whole-number part, we <u>compare the fraction parts to decide which is larger</u>. For example:

    $10\frac{5}{8}$ is larger than $10\frac{3}{8}$, since $\frac{5}{8}$ is larger than $\frac{3}{8}$

    Draw a circle around the <u>larger</u> mixed number in each pair.   (a) $3\frac{1}{6}$ and $3\frac{5}{6}$ (b) $15\frac{8}{9}$ and $15\frac{5}{9}$

    a) $3\frac{3}{4}$ b) $35\frac{1}{7}$

82. When two mixed numbers with the same whole-number parts have fraction parts with "unlike" denominators, <u>we have to get common denominators</u> before we can compare their size. For example:

    $4\frac{5}{6}$ is larger than $4\frac{2}{3}$, since $\frac{5}{6}$ is larger than $\frac{2}{3}$ $\left(\text{or } \frac{4}{6}\right)$

    $7\frac{4}{5}$ is larger than $7\frac{3}{4}$, since $\frac{4}{5}$ $\left(\text{or } \frac{16}{20}\right)$ is larger than $\frac{3}{4}$ $\left(\text{or } \frac{15}{20}\right)$

    Draw a circle around the <u>larger</u> mixed number in each pair.   (a) $9\frac{1}{3}$ or $9\frac{1}{4}$ (b) $1\frac{9}{16}$ or $1\frac{5}{8}$

    a) $3\frac{5}{6}$ b) $15\frac{8}{9}$

83. Of the mixed numbers: $15\frac{1}{4}$, $14\frac{7}{8}$, $15\frac{3}{8}$, $14\frac{3}{4}$, $15\frac{1}{8}$

    (a) The largest is _____.   (b) The smallest is _____.

    a) $9\frac{1}{3}$ b) $1\frac{5}{8}$

    a) $15\frac{3}{8}$ b) $14\frac{3}{4}$

5-12 APPLIED PROBLEMS

This section contains some verbal or applied problems. All of them can be solved by a subtraction involving mixed numbers or by comparing the size of mixed numbers.

84. While on a diet, a man's weight dropped from $196\frac{1}{4}$ pounds to $174\frac{3}{4}$ pounds. How many pounds did he lose?

85. The baggage allowance on an airplane is 42 pounds. If a woman's suitcase weighs $37\frac{1}{4}$ pounds, how much underweight is it?

$21\frac{1}{2}$ pounds

$4\frac{3}{4}$ pounds

86. Jim is $59\frac{1}{4}$ inches tall. Paul is $58\frac{1}{2}$ inches tall.

   (a) Which boy is taller?

   (b) How much taller is he?

87. To drive the same 100 miles, a first car used $6\frac{1}{2}$ gallons of gasoline and a second car used $6\frac{7}{10}$ gallons of gasoline.

   (a) Which car used more gasoline?

   (b) How much more did it use?

---

a) Jim
b) $\frac{3}{4}$ inch

a) second car
b) $\frac{1}{5}$ gallon

SELF-TEST 12 (Frames 43-87)

Do the following subtractions.

1. $4\frac{5}{6} - 3\frac{1}{2} =$

2. $5\frac{1}{8} - 4\frac{7}{8} =$

3. $8\frac{1}{4} - 2\frac{9}{10} =$

4. $1\frac{3}{5} - \frac{7}{10} =$

5. $12 - 5\frac{3}{8} =$

6. $1 - \frac{1}{6} =$

ANSWERS: 1. $1\frac{1}{3}$ 2. $\frac{1}{4}$ 3. $5\frac{7}{20}$ 4. $\frac{9}{10}$ 5. $6\frac{5}{8}$ 6. $\frac{5}{6}$

5-13 MULTIPLICATIONS INVOLVING MIXED NUMBERS

In this section, we will discuss multiplications in which the factors are either two mixed numbers, a whole number and a mixed number, or a proper fraction and a mixed number. As a general strategy for these multiplications, we will convert the mixed numbers to improper fractions.

88. In the multiplication at the right, both factors are mixed numbers. To perform the multiplication, we use the following steps:

    $1\frac{1}{4} \times 3\frac{1}{2}$

    (1) Convert both mixed numbers to <u>improper</u> fractions.

    $1\frac{1}{4} \times 3\frac{1}{2} = \frac{5}{4} \times \frac{7}{2}$

    (2) Perform the multiplication of improper fractions in the usual way.

    $1\frac{1}{4} \times 3\frac{1}{2} = \frac{5}{4} \times \frac{7}{2} = \frac{35}{8}$

    (3) Convert the product back to a mixed number.

    $1\frac{1}{4} \times 3\frac{1}{2} = \frac{5}{4} \times \frac{7}{2} = \frac{35}{8} = 4\frac{3}{8}$

    Following the steps above, complete this multiplication. $1\frac{1}{2} \times 3\frac{3}{4} = \underline{\quad} \times \underline{\quad} = \underline{\quad} = \underline{\quad}$

89. Before multiplying the two improper fractions, look for opportunities to cancel. An example is given below. Complete the other problem.

    $\frac{3}{2} \times \frac{15}{4} = \frac{45}{8} = 5\frac{5}{8}$

    $2\frac{1}{4} \times 1\frac{2}{3} = \frac{\cancel{9}^{3}}{4} \times \frac{5}{\cancel{3}_{1}} = \frac{15}{4} = 3\frac{3}{4}$ $1\frac{2}{3} \times 2\frac{1}{5} = \underline{\quad} \times \underline{\quad} = \underline{\quad} = \underline{\quad}$

90. When multiplying two mixed numbers, the product can be a whole number. An example is given below. Complete the other problem.

    $\frac{\cancel{2}^{1}}{3} \times \frac{11}{\cancel{2}_{1}} = \frac{11}{3} = 3\frac{2}{3}$

    $2\frac{2}{5} \times 1\frac{1}{4} = \frac{\cancel{12}^{3}}{\cancel{5}_{1}} \times \frac{\cancel{5}^{1}}{\cancel{4}_{1}} = \frac{3}{1} = 3$ $1\frac{2}{3} \times 3\frac{3}{5} = \underline{\quad} \times \underline{\quad} = \underline{\quad}$

91. To multiply a mixed number and a whole number, we simply convert the mixed number to an improper fraction. For example:

    $\frac{\cancel{5}^{1}}{\cancel{3}_{1}} \times \frac{\cancel{18}^{6}}{\cancel{3}_{1}} = \frac{6}{1} = 6$

    $3 \times 2\frac{1}{4} = 3 \times \frac{9}{4} = \frac{27}{4} = 6\frac{3}{4}$ $4\frac{3}{7} \times 2 = \underline{\quad} \times \underline{\quad} = \underline{\quad} = \underline{\quad}$

92. When multiplying a mixed number and a whole number, be sure to cancel wherever possible. For example:

    $\frac{31}{7} \times 2 = \frac{62}{7} = 8\frac{6}{7}$

    $3 \times 5\frac{5}{6} = \cancel{3}^{1} \times \frac{35}{\cancel{6}_{2}} = \frac{35}{2} = 17\frac{1}{2}$ $2\frac{1}{7} \times 14 = \underline{\quad} \times \underline{\quad} = \underline{\quad}$

93. To multiply a mixed number and a proper fraction, we simply convert the mixed number to an improper fraction. For example:

    $\frac{15}{\cancel{7}_{1}} \times \cancel{14}^{2} = 30$

    $\frac{1}{2} \times 5\frac{2}{3} = \frac{1}{2} \times \frac{17}{3} = \frac{17}{6} = 2\frac{5}{6}$ $1\frac{4}{5} \times \frac{3}{4} = \underline{\quad} \times \underline{\quad} = \underline{\quad} = \underline{\quad}$

112

94. When multiplying a mixed number and a proper fraction, also cancel wherever possible. For example:

$$6\frac{3}{10} \times \frac{2}{3} = \frac{\overset{21}{\cancel{63}}}{\underset{5}{\cancel{10}}} \times \frac{\overset{1}{\cancel{2}}}{\underset{1}{\cancel{3}}} = \frac{21}{5} = 4\frac{1}{5} \qquad \frac{4}{7} \times 2\frac{5}{8} = \underline{\hspace{3cm}}$$

$\left| \dfrac{9}{5} \times \dfrac{3}{4} = \dfrac{27}{20} = 1\dfrac{7}{20} \right.$

95. When mixed numbers are converted to improper fractions, the numerators of the improper fractions can be large. In such cases, we have to perform a long multiplication to find the numerator of the product. An example is given below.

In the multiplication at the right, we got 713 by multiplying 31 by 23.

$$11\frac{1}{2} \times 10\frac{1}{3} = \frac{23}{2} \times \frac{31}{3} = \frac{713}{6}$$

Since $\dfrac{713}{6} = 6\overline{)713}^{\,118\text{ r}5}$, $11\dfrac{1}{2} \times 10\dfrac{1}{3} = \underline{\hspace{2cm}}$.

$\left| 1\dfrac{1}{2} \right.$

96. When multiplying two mixed numbers, the denominator of the improper-fraction form of the product can also be large. For example:

$$7\frac{1}{4} \times 5\frac{7}{8} = \frac{29}{4} \times \frac{47}{8} = \frac{1,363}{32}$$

To convert the product back to a mixed number, we have to perform a long division as we have done at the right.

Therefore: $7\dfrac{1}{4} \times 5\dfrac{7}{8} = \underline{\hspace{2cm}}$

$\left| 118\dfrac{5}{6} \right.$

```
      4 2 r19
32)1,3 6 3
    1 2 8
      8 3
      6 4
      1 9
```

$\left| 42\dfrac{19}{32} \right.$

5-14   THREE-FACTOR MULTIPLICATIONS INVOLVING MIXED NUMBERS

Three-factor multiplications involving mixed numbers are also performed by converting the mixed numbers to improper fractions. We will discuss the procedure in this section.

97. The multiplication below involves three mixed numbers. To perform it, we converted the mixed numbers to improper fractions and then proceeded in the usual way.

$$2\frac{1}{2} \times 2\frac{1}{3} \times 1\frac{1}{4} = \frac{5}{2} \times \frac{7}{3} \times \frac{5}{4} = \frac{175}{24} = 7\frac{7}{24}$$

Note: To convert $\dfrac{175}{24}$ back to a mixed number, we had to divide 175 by 24.

Following the same steps, complete this multiplication. $1\dfrac{2}{5} \times 1\dfrac{1}{2} \times 1\dfrac{3}{4} = \underline{\hspace{3cm}}$

$\left| 3\dfrac{27}{40} \text{ , from } \dfrac{147}{40} \right.$

98. When multiplying three mixed numbers, look for opportunities to cancel in order to simplify the work. For example:

$$1\frac{1}{2} \times 2\frac{1}{3} \times 1\frac{1}{4} = \frac{\cancel{3}^{1}}{2} \times \frac{7}{\cancel{3}_{1}} \times \frac{5}{4} = \frac{35}{8} = 4\frac{3}{8}$$

Use the same steps to complete this multiplication. $1\frac{2}{5} \times 2\frac{1}{2} \times 3\frac{1}{3} =$ _____

99. The multiplication below contains a mixed number together with a whole number and a proper fraction. Notice that we converted the mixed number to an improper fraction and then proceeded in the usual way.

$11\frac{2}{3}$, from $\frac{7}{\cancel{5}_{1}} \times \frac{\cancel{5}^{1}}{\cancel{2}_{1}} \times \frac{\cancel{10}^{5}}{3} = \frac{35}{3}$

$$5 \times 2\frac{1}{2} \times \frac{3}{4} = 5 \times \frac{5}{2} \times \frac{3}{4} = \frac{5 \times 5 \times 3}{2 \times 4} = \frac{75}{8} = 9\frac{3}{8}$$

Following the steps above, complete this multiplication. Cancel wherever possible. $1\frac{3}{5} \times 20 \times \frac{3}{7} =$ _____

100. In the multiplication below, the numerators of the three improper fractions are large.

$13\frac{5}{7}$, from $\frac{8}{\cancel{5}_{1}} \times \cancel{20}^{4} \times \frac{3}{7} = \frac{96}{7}$

$$12\frac{1}{2} \times 10\frac{1}{2} \times 7\frac{3}{4} = \frac{25}{2} \times \frac{21}{2} \times \frac{31}{4} = \frac{16,275}{16}$$

To find the numerator of the product, we had to perform two long multiplications. That is:

(1) We multiplied 25 and 21 and got 525.

(2) Then we multiplied 525 and 31 and got 16,275.

To complete the multiplication, we must divide 16,275 by 16 in order to convert the product back to a mixed number.

Since $16 \overline{)16,275}^{\,1,017 \text{ r}3}$, $12\frac{1}{2} \times 10\frac{1}{2} \times 7\frac{3}{4} =$ _____

$1,017\frac{3}{16}$

5-15   DIVISIONS INVOLVING MIXED NUMBERS

To perform divisions involving mixed numbers, we simply convert the mixed numbers to improper fractions and then proceed in the usual way. We will discuss the procedure in this section.

101. Divisions involving mixed numbers are also performed by converting them to multiplications. Before doing so, however, we must convert the mixed numbers to improper fractions. For example:

$$1\frac{2}{3} \div 1\frac{3}{4} = \frac{5}{3} \div \frac{7}{4} = \frac{5}{3} \times \left(\text{the reciprocal of } \frac{7}{4}\right) = \frac{5}{3} \times \frac{4}{7} = \frac{20}{21}$$

Note: To convert the division to a multiplication, we multiplied the dividend $\frac{5}{3}$ by the reciprocal of the divisor.

Following the steps above, complete the division below. Write the quotient in lowest terms.

$$3\frac{1}{3} \div 1\frac{1}{9} = \frac{10}{3} \div \frac{10}{9} = \frac{10}{3} \times \left(\text{the reciprocal of } \frac{10}{9}\right) = \underline{\qquad} \times \underline{\qquad} = \underline{\qquad}$$

114

102. When the division involves a proper fraction and a mixed number, we also convert the mixed number to an improper fraction <u>before</u> converting the division to a multiplication. For example:

$$\frac{3}{4} \div 1\frac{3}{7} = \frac{3}{4} \div \frac{10}{7} = \frac{3}{4} \times \frac{7}{10} = \frac{21}{40} \qquad 5\frac{1}{2} \div \frac{1}{3} = \frac{11}{2} \div \frac{1}{3} = \underline{} \times \underline{} = \underline{}$$

$$\frac{\cancel{10}^{1}}{\cancel{3}_{1}} \times \frac{\cancel{9}^{3}}{\cancel{10}_{1}} = 3$$

103. When the division involves a whole number and a mixed number, we also convert the mixed number to an improper fraction <u>before</u> converting the division to a multiplication. For example:

$$3\frac{1}{4} \div 5 = \frac{13}{4} \div 5 = \frac{13}{4} \times \frac{1}{5} = \frac{13}{20} \qquad 10 \div 3\frac{1}{2} = 10 \div \frac{7}{2} = \underline{} \times \underline{} = \underline{}$$

$$\frac{11}{2} \times 3 = 16\frac{1}{2}$$

104. Divisions involving mixed numbers are frequently written in complex-fraction form. For example:

$$7\frac{1}{3} \div 6\frac{1}{8} \text{ is written } \frac{7\frac{1}{3}}{6\frac{1}{8}} \qquad 3 \div 5\frac{5}{6} \text{ is written } \underline{}$$

$$10 \times \frac{2}{7} = 2\frac{6}{7}$$

105. Just as $\dfrac{4\frac{3}{8}}{7\frac{5}{16}}$ means "divide $4\frac{3}{8}$ by $7\frac{5}{16}$", $\dfrac{12\frac{3}{4}}{7}$ means "divide ____ by ____".

$$\frac{3}{5\frac{5}{6}}$$

106. To perform divisions involving mixed numbers when they are written in complex-fraction form, we use the same steps. That is, we convert the mixed numbers to improper fractions and then convert the division to a multiplication. For example:

$$12\frac{3}{4} \text{ by } 7$$

$$\frac{3\frac{1}{2}}{1\frac{3}{5}} = \frac{\frac{7}{2}}{\frac{8}{5}} = \frac{7}{2} \times \left(\text{the reciprocal of } \frac{8}{5}\right) = \frac{7}{2} \times \frac{5}{8} = \frac{35}{16} = 2\frac{3}{16}$$

$$\frac{3\frac{2}{3}}{5} = \frac{\frac{11}{3}}{5} = \frac{11}{3} \times (\text{the reciprocal of } 5) = \frac{11}{3} \times \frac{1}{5} = \frac{11}{15}$$

$$\frac{3}{1\frac{3}{4}} = \frac{3}{\frac{7}{4}} = 3 \times \left(\text{the reciprocal of } \frac{7}{4}\right) = \underline{} \times \underline{} = \underline{}$$

107. Complete each division below. Write each quotient in lowest terms.

$$3 \times \frac{4}{7} = \frac{12}{7} = 1\frac{5}{7}$$

(a) $\dfrac{10\frac{2}{3}}{12\frac{4}{5}} =$

(b) $\dfrac{2\frac{1}{4}}{9} =$

(c) $\dfrac{6}{1\frac{5}{7}} =$

108. When dividing mixed numbers, the terms involved can be quite large. For example:

a) $\frac{5}{6}$, from $\frac{\cancel{32}^{1}}{3} \times \frac{5}{\cancel{64}_{2}}$ b) $\frac{1}{4}$, from $\frac{\cancel{8}^{1}}{4} \times \frac{1}{\cancel{8}_{1}}$ c) $3\frac{1}{2}$, from $\frac{1}{\cancel{6}} \times \frac{7}{\cancel{12}_{2}}$

$$\frac{42\frac{1}{2}}{10\frac{6}{7}} = \frac{\frac{85}{2}}{\frac{76}{7}} = \frac{85}{2} \times \frac{7}{76} = \frac{595}{152} = 3\frac{139}{152}$$

$$\frac{5\frac{7}{32}}{7\frac{9}{16}} = \frac{\frac{167}{32}}{\frac{121}{16}} = \frac{167}{32} \times \frac{16}{121} = \underline{\qquad}$$

$\frac{167}{242}$, from $\frac{167}{\cancel{32}_{2}} \times \frac{\cancel{16}^{1}}{121}$

5-16 MULTIPLICATIONS AND DIVISIONS INVOLVING A MIXED NUMBER AND EITHER "1" OR "0"

In this section, we will briefly discuss multiplications and divisions involving a mixed number and either "1" or "0". Divisions of a mixed number by itself are also included.

109. Whenever we <u>multiply</u> a <u>mixed number</u> and "1", the <u>product is identical to the mixed number</u>. For example:

$1 \times 2\frac{3}{5} = 1 \times \frac{13}{5} = \frac{13}{5} = 2\frac{3}{5}$ (a) $1 \times 10\frac{7}{8} =$ _____   (b) $42\frac{1}{3} \times 1 =$ _____

110. Whenever we <u>multiply</u> a <u>mixed number</u> and "0", the <u>product is</u> "0". For example:

a) $10\frac{7}{8}$ b) $42\frac{1}{3}$

$0 \times 7\frac{3}{4} = 0 \times \frac{31}{4} = 0$ (a) $0 \times 2\frac{5}{16} =$ _____   (b) $1\frac{7}{8} \times 0 =$ _____

111. Whenever we <u>divide</u> a <u>mixed number</u> <u>by itself</u>, the <u>quotient is</u> "1". For example:

a) 0   b) 0

$3\frac{1}{2} \div 3\frac{1}{2} = \frac{7}{2} \div \frac{7}{2} = \frac{7}{2} \times \frac{2}{7} = 1$ $\frac{1\frac{2}{3}}{1\frac{2}{3}} = \frac{\frac{5}{3}}{\frac{5}{3}} = \frac{5}{3} \times \frac{3}{5} = 1$

Using the fact above, write each quotient at the right.

(a) $\frac{33\frac{7}{8}}{33\frac{7}{8}} =$ _____   (b) $10\frac{1}{2} \div 10\frac{1}{2} =$ _____

112. Whenever we <u>divide</u> a <u>mixed number</u> <u>by</u> "1", the <u>quotient is the mixed number</u>. For example:

a) 1   b) 1

$2\frac{3}{4} \div 1 = \frac{11}{4} \div 1 = \frac{11}{4} \times 1 = \frac{11}{4} = 2\frac{3}{4}$ $\frac{5\frac{1}{3}}{1} = \frac{\frac{16}{3}}{1} = \frac{16}{3} \times 1 = \frac{16}{3} = 5\frac{1}{3}$

Note: The reciprocal of "1" is "1", since $1 \times 1 = 1$.

Using the fact above, write each quotient at the right.

(a) $12\frac{3}{8} \div 1 =$ _____   (b) $\frac{35\frac{3}{16}}{1} =$ _____

115

116

113. We have performed a division of "1" by a mixed number below. Complete the other division.

$$\frac{1}{1\frac{3}{5}} = \frac{1}{\frac{8}{5}} = 1 \times \frac{5}{8} = \frac{5}{8} \qquad\qquad 1 \div 3\frac{1}{7} = 1 \div \frac{22}{7} = 1 \times \underline{\qquad} = \underline{\qquad}$$

a) $12\frac{3}{8}$ b) $35\frac{3}{16}$

114. Whenever we divide "0" by a mixed number, the quotient is "0". For example:

$$0 \div 5\frac{1}{8} = 0 \div \frac{41}{8} = 0 \times \frac{8}{41} = 0 \qquad\qquad \frac{0}{2\frac{5}{6}} = \frac{0}{\frac{17}{6}} = 0 \times \frac{6}{17} = 0$$

Using the fact above, write each quotient at the right.

(a) $\dfrac{0}{3\frac{11}{16}} = \underline{\qquad}$

(b) $0 \div 100\frac{3}{4} = \underline{\qquad}$

$1 \times \frac{7}{22} = \frac{7}{22}$

115. Complete: (a) $1 \times 8\frac{1}{2} = \underline{\qquad}$ (b) $8\frac{1}{2} \div 1 = \underline{\qquad}$ (c) $8\frac{1}{2} \times 0 = \underline{\qquad}$

a) 0   b) 0

116. Complete: (a) $\dfrac{6\frac{2}{3}}{6\frac{2}{3}} = \underline{\qquad}$ (b) $\dfrac{1}{6\frac{2}{3}} = \underline{\qquad}$ (c) $\dfrac{0}{6\frac{2}{3}} = \underline{\qquad}$

a) $8\frac{1}{2}$ b) $8\frac{1}{2}$ c) 0

a) 1   b) $\frac{3}{20}$ c) 0

5-17   CONTRASTING THE FOUR OPERATIONS WITH MIXED NUMBERS

There are four basic operations with mixed numbers: addition, subtraction, multiplication, and division. In this section, we will briefly contrast the processes for those four operations.

117. To perform additions and subtractions involving mixed numbers, we have used the "whole number - fraction" method. For example:

$$5\frac{1}{7} + 3\frac{4}{7} = (5 + 3) + \left(\frac{1}{7} + \frac{4}{7}\right) = 8\frac{5}{7} \qquad\qquad 7\frac{8}{9} - 1\frac{1}{9} = (7 - 1) + \left(\frac{8}{9} - \frac{1}{9}\right) = \underline{\qquad}$$

118. To perform multiplications and divisions involving mixed numbers, we have converted the mixed numbers to improper fractions. For example:

$$1\frac{2}{3} \times 3\frac{1}{2} = \frac{5}{3} \times \frac{7}{2} = \frac{35}{6} = 5\frac{5}{6} \qquad\qquad 3\frac{1}{5} \div 1\frac{1}{2} = \frac{16}{5} \div \frac{3}{2} = \underline{\qquad} \times \underline{\qquad} = \underline{\qquad}$$

$6\frac{7}{9}$

119. In which operations below would we have to convert the mixed numbers to improper fractions? $\underline{\qquad}$

(a) $3\frac{1}{2} + 1\frac{2}{3}$ (b) $3\frac{1}{2} \times 1\frac{2}{3}$ (c) $3\frac{1}{2} - 1\frac{2}{3}$ (d) $3\frac{1}{2} \div 1\frac{2}{3}$

$\dfrac{16}{5} \times \dfrac{2}{3} = 2\dfrac{2}{15}$

Only (b) and (d)

120. In which operations below would we use the "whole number - fraction" method without converting the mixed numbers to improper fractions?

(a) $5\frac{1}{5} + 1\frac{1}{2}$ (b) $5\frac{1}{5} \times 1\frac{1}{2}$ (c) $5\frac{1}{5} - 1\frac{1}{2}$ (d) $5\frac{1}{5} \div 1\frac{1}{2}$ | Only (a) and (b)

5-18   ESTIMATING ANSWERS TO MIXED-NUMBER PROBLEMS

In this section, we will briefly discuss some methods for estimating answers to problems involving mixed numbers. This estimation procedure is helpful since it can be used to avoid large errors.

121. When performing an addition or subtraction involving mixed numbers, we can estimate the approximate answer by simply ignoring the fraction part of the mixed numbers. For example:

The answer for $7\frac{1}{2} + 9\frac{1}{8}$ is approximately $7 + 9$ or 16.

The answer for $12\frac{3}{4} - 8\frac{5}{16}$ is approximately $12 - 8$ or 4.

By estimating an approximate answer, we can check the sensibleness of an obtained answer. For example:

For $5\frac{1}{9} + 10\frac{2}{9}$, a student got the answer $15\frac{1}{3}$. Since the approximate answer is $5 + 10$ or 15, his answer makes sense.

For $25\frac{3}{5} - 7\frac{1}{5}$, a student got the answer $8\frac{2}{5}$. Since the approximate answer is $25 - 7$ or 18, his answer does not make sense.

Of the possible answers: $10\frac{1}{4}$, $12\frac{1}{4}$, $14\frac{1}{4}$, $16\frac{1}{4}$, $18\frac{1}{4}$

(a) Which one makes sense for $7\frac{1}{8} + 5\frac{1}{8}$? _____   (b) Which one makes sense for $20\frac{1}{2} - 4\frac{1}{4}$? _____

122. When performing a multiplication involving mixed numbers, we can also estimate the approximate answer by simply ignoring the fraction part of the mixed numbers. | a) $12\frac{1}{4}$ b) $16\frac{1}{4}$

The approximate answer can be used to check the sensibleness of the obtained answer. For example:

For $3\frac{1}{3} \times 2\frac{1}{4}$, a student got the answer $7\frac{1}{2}$. Since the approximate answer is 3×2 or 6, his answer makes sense.

For $5\frac{1}{2} \times 3\frac{1}{4}$, a student got the answer $9\frac{7}{8}$. Since the approximate answer is 5×3 or 15, his answer does not make sense.

Of the possible answers: $7\frac{5}{12}$, $14\frac{1}{4}$, $22\frac{1}{4}$, $33\frac{19}{32}$, $45\frac{19}{32}$

(a) Which one makes sense for $3\frac{1}{6} \times 4\frac{1}{2}$? _____   (b) Which one makes sense for $5\frac{3}{8} \times 6\frac{1}{4}$? _____

| a) $14\frac{1}{4}$ b) $33\frac{19}{32}$

123. We can use the same method to estimate the answer for a division involving mixed numbers. The approximate answer can be used to check the sensibleness of the obtained answer. For example:

For $8\frac{1}{2} \div 2\frac{1}{4}$, a student got the answer $3\frac{7}{9}$. Since the approximate answer is $8 \div 2$ or 4, his answer <u>makes sense</u>.

For $6\frac{1}{4} \div 3\frac{3}{8}$, a student got the answer $\frac{5}{27}$. Since the approximate answer is $6 \div 3$ or 2, his answer <u>does not make sense</u>.

Of the possible answers: $1\frac{1}{10}$, $2\frac{21}{26}$, $5\frac{15}{16}$, $8\frac{5}{32}$, $10\frac{5}{8}$

(a) Which one makes sense for $9\frac{1}{8} \div 3\frac{1}{4}$? _____ (b) Which one makes sense for $2\frac{3}{4} \div 2\frac{1}{2}$? _____

a) $2\frac{21}{26}$ b) $1\frac{1}{10}$

5-19   MIXED APPLIED PROBLEMS

The section contains some verbal or applied problems. After two problems requiring a multiplication and two requiring a division involving mixed numbers, we will give some mixed problems in which all four of the operations involving mixed numbers are used.

---

124. A student worked $3\frac{1}{4}$ hours per day for each of 5 days. Find the total number of hours worked.

(Think: $5 \times 3\frac{1}{4}$)

$16\frac{1}{4}$ hours

125. A woman used $4\frac{1}{2}$ gallons of gasoline on a trip. If her car averaged $12\frac{1}{2}$ miles per gallon, how long was the trip?

(Think: $4\frac{1}{2} \times 12\frac{1}{2}$)

$56\frac{1}{4}$ miles

126. A board is $7\frac{3}{4}$ feet long. If it is divided into 3 equal parts, how long is each part?

(Think: $7\frac{3}{4} \div 3$)

$2\frac{7}{12}$ feet

127. A man drove 210 miles in $3\frac{1}{2}$ hours. What was his average speed in miles per hour?

(Think: $210 \div 3\frac{1}{2}$)

60 miles per hour

128. A man bought two adjoining lots. One contained $3\frac{5}{8}$ acres; the other contained $5\frac{3}{4}$ acres. He later sold $6\frac{1}{2}$ acres of the land. To find out how many acres he had left, answer the questions below.

(a) Find the total number of acres bought.

(b) Find the number of acres left after he sold $6\frac{1}{2}$ acres.

a) $9\frac{3}{8}$ acres   b) $2\frac{7}{8}$ acres

129. Joe walked to school and back 20 days in one month. If the round-trip distance is $1\frac{1}{8}$ miles, how far did he walk?

$22\frac{1}{2}$ miles

130. How many pieces of rope $1\frac{3}{4}$ feet long can be cut from a rope that is $10\frac{1}{2}$ feet long?

6 pieces

SELF-TEST 13 (Frames 88-130)

Do the following problems.

1. $3\frac{3}{4} \times 1\frac{1}{3} =$

2. $7\frac{1}{2} \div 30 =$

3. $4 \times 2\frac{3}{8} =$

4. $5\frac{1}{4} \div 2\frac{1}{3} =$

5. $21 \div 4\frac{2}{3} =$

6. $2 \times 1\frac{1}{3} \times 2\frac{3}{4} =$

7. $\dfrac{5\frac{2}{3}}{1} =$

8. $6\frac{3}{8} \times 1 =$

9. $\dfrac{9\frac{3}{4}}{9\frac{3}{4}} =$

10. $3\frac{4}{5} \times 0 =$

ANSWERS: 1. 5   3. $9\frac{1}{2}$ 5. $4\frac{1}{2}$ 7. $5\frac{2}{3}$ 9. 1

2. $\frac{1}{4}$ 4. $2\frac{1}{4}$ 6. $7\frac{1}{3}$ 8. $6\frac{3}{8}$ 10. 0

119

Unit 6 PRIME FACTORING AND LOWEST COMMON DENOMINATORS

Up to this point, we have used the "direct" method to identify lowest common denominators. However, there are cases in which the "direct" method is quite difficult to use. In such cases, we can use another method called the "prime factoring" method. In this unit, we will discuss "prime factoring" and the "prime factoring" method for identifying lowest common denominators.

6-1 THE NEED FOR ANOTHER METHOD FOR IDENTIFYING LOWEST COMMON DENOMINATORS

In this section, we will show that the "direct" method for identifying lowest common denominators is quite difficult to use in some cases.

1. When using the "direct" method to identify a lowest common denominator, we simply check multiples of the larger denominator until we find the lowest one that is also a multiple of the smaller denominator. An example is given below.

   To find the lowest common denominator for $\frac{3}{10} + \frac{5}{14}$, we check the multiples of 14 until we find the lowest one that is also a multiple of 10.

       Is 14 a multiple of 10? No
       Is 28 a multiple of 10? No
       Is 42 a multiple of 10? No
       Is 56 a multiple of 10? No
       Is 70 a multiple of 10? Yes

   Therefore, the lowest common denominator for $\frac{3}{10} + \frac{5}{14}$ is _____.

2. We have also used the "direct" method for additions of three fractions. An example is given below.

   To find the lowest common denominator for $\frac{1}{2} + \frac{3}{7} + \frac{1}{4}$, we check the multiples of 7 until we find the lowest one that is also a multiple of both 2 and 4.

       Is 7 a multiple of both 2 and 4? No
       Is 14 a multiple of both 2 and 4? No
       Is 21 a multiple of both 2 and 4? No
       Is 28 a multiple of both 2 and 4? Yes

   Therefore, the lowest common denominator for $\frac{1}{2} + \frac{3}{7} + \frac{1}{4}$ is _____.

Answer: 70

3. Sometimes the "direct" method for identifying a lowest common denominator is difficult to use. An example is given below.

To find the lowest common denominator for $\frac{5}{18} + \frac{1}{14}$, we check multiples of 18 until we find the lowest one that is also a multiple of 14.

Is 18 a multiple of 14? No
Is 36 a multiple of 14? No
Is 54 a multiple of 14? No
Is 72 a multiple of 14? No
Is 90 a multiple of 14? No
Is 108 a multiple of 14? No
Is 126 a multiple of 14? Yes

Though we found the lowest common denominator by the "direct" method, it was difficult to list the multiples of 18 and to check whether each one was also a multiple of 14.

4. Here is another case in which the "direct" method is difficult to use.

To find the lowest common denominator for $\frac{1}{20} + \frac{3}{16} + \frac{7}{12}$, we check multiples of 20 until we find the lowest one that is also a multiple of both 16 and 12.

Is 20 a multiple of both 16 and 12? No
Is 40 a multiple of both 16 and 12? No
Is 60 a multiple of both 16 and 12? No
Is 80 a multiple of both 16 and 12? No
Is 100 a multiple of both 16 and 12? No
Is 120 a multiple of both 16 and 12? Yes

Though we found the lowest common denominator by the "direct" method, it was difficult to check whether each multiple of 20 was also a multiple of both 16 and 12.

Though the "direct" method for identifying lowest common denominators works in all cases, there are cases in which it is difficult to use. In such cases, another method called the "prime factoring" method can be used. The purpose of this unit is to discuss "prime" factoring" and the "prime factoring" method for identifying lowest common denominators.

6-2 FACTORS AND THE FACTORING PROCESS

In this section, we will discuss what is meant by factors and the factoring process. The discussion will be limited to the whole-number factors of whole numbers.

5. Factoring is a process in which a number is written as a multiplication. That is:

14 can be factored into 2 x 7   30 can be factored into 6 x 5

When the factoring process is used, the pair of numbers in the multiplication are called "factors". Write the missing factor in each blank below.

(a) 10 = 2 x ____   (b) 49 = ____ x 7   (c) 15 = 5 x ____

a) 2 x <u>5</u>   b) <u>7</u> x 7   c) 5 x <u>3</u>

122

6. One pair of factors for any number is the number itself and "1". That is:

    3 = 1 x 3   7 = 1 x 7   12 = 1 x 12   99 = 1 x ____

7. To check whether the factoring of a number is correct, we simply perform the multiplication. For example:

    The factoring 63 = 7 x 9 is correct, since 7 x 9 = 63.

    The factoring 24 = 4 x 8 is <u>not</u> correct, since 4 x 8 = 32.

Which of the following factorings are correct? _____

    (a) 56 = 8 x 7   (b) 54 = 5 x 9   (c) 72 = 8 x 8   (d) 42 = 3 x 14

> 1 x <u>99</u>

8. Many numbers can be factored into various pairs of factors. For example:

The number "18" can be factored into three pairs of factors. We have done so below.	(a) The number "6" can be factored into two pairs of factors. Do so below.	(b) The number "16" can be factored into three pairs of factors. Do so below.
18 = 1 x 18	6 = ____ x ____	16 = ____ x ____
18 = 2 x 9	6 = ____ x ____	16 = ____ x ____
18 = 3 x 6		16 = ____ x ____

> Only (a) and (d)

9. A first number is a factor of a second number if the second number is <u>divisible</u> by the first number. For example:

    3 is a factor of 15, since 15 is divisible by 3.
    12 is a factor of 24, since 24 is divisible by 12.

Which of the following numbers are factors of 36? _____

    (a) 7   (b) 2   (c) 9   (d) 8   (e) 18   (f) 11

> a) 1 x 6   b) 1 x 16
>   2 x 3   2 x 8
>   4 x 4

10. It should be obvious that a number can have many factors. For example:

The number "12" has <u>six</u> factors, since 12 is divisible by 1, 2, 3, 4, 6, and 12.

The number "10" has <u>four</u> factors, since 10 is divisible by ____, ____, ____, and ____.

> (b), (c), and (e)

11. (a) The four factors of "14" are ____, ____, ____, and ____.

    (b) The six factors of "20" are ____, ____, ____, ____, ____, and ____.

> 1, 2, 5, and 10

12. Some numbers can be written as a multiplication containing three or more factors.

    For example: 12 = 2 x 2 x 3   48 = 2 x 2 x 3 x 4

Complete each factoring: (a) 36 = 2 x 3 x ____   (b) 24 = 2 x 2 x 2 x ____

> a) 1, 2, 7, 14
> b) 1, 2, 4, 5, 10, 20

> a) 6   b) 3

13. We can also check a factoring containing three or more factors by performing the multiplication. Remember that multiplications of that type are performed "two at a time", and that they can be performed either from left to right or from right to left. An example is given below.

    To check whether the factoring 90 = 2 x 3 x 3 x 5 is correct, we have performed the multiplication two different ways below.

    ```
    |2 x 3| x 3 x 5           2 x 3 x |3 x 5|
         ↓                             ↓
    | 6  x 3| x 5             2 x |3 x  15|
         ↓                             ↓
      18     x 5 = 90          2 x   45   = 90
    ```

    Is 90 = 2 x 3 x 3 x 5 a correct factoring? _____

---

14. (a) Is the factoring below correct? ____   (b) Is the factoring below correct? ____

    105 = 3 x 5 x 7   110 = 2 x 2 x 2 x 3 x 5

| Yes |

| a) Yes, since 3 x 5 x 7 = 105 |
| b) No, since 2 x 2 x 2 x 3 x 5 = 120 |

6-3   PRIME AND COMPOSITE NUMBERS

In this section, we will discuss the difference between prime and composite numbers. Then in the next section, we will discuss what is meant by the "prime factoring" of a number.

15. Some numbers have only "1" and themselves as factors. For example:

    The only factors of "2" are 1 and 2.   The only factors of "7" are 1 and 7.

    Most numbers, however, have one or more factors other than themselves and "1". For example:

    "4" has 1, 4, and 2 as its factors.   "15" has 1, 15, 3 and 5 as its factors.

    (a) Does "5" have any factor other than 1 and 5? _____

    (b) Does "9" have any factor other than 1 and 9? _____

    (c) Does "12" have any factor other than 1 and 12? _____

---

16. Any number larger than "1" that has only itself and "1" as its factors is called a "prime" number. For example:

    11 is a "prime" number since its only factors are 1 and 11.

    Which of the following are "prime" numbers? _____

    (a) 3   (b) 8   (c) 10   (d) 13   (e) 16

| a) No. |
| b) Yes. It has 3. |
| c) Yes. It has 2, 3, 4, and 6. |

---

17. Any number larger than "1" that has at least one factor other than itself and "1" is called a "composite" number. For example:

    9 is a "composite" number since it has 3 as a factor.

    14 is a "composite" number since it has 2 and 7 as factors.

    Which of the following are "composite" numbers? _____

    (a) 4   (b) 7   (c) 12   (d) 18   (e) 23

| Only (a) and (d) |

123

18. Only numbers larger than "1" are called "prime" or "composite" numbers. The number "1" is neither a prime nor a composite number.

    Is the number "1" a prime or a composite number? _____

    | Neither

19. The number "2" is a prime number since its only factors are 1 and 2. However, all other "even" numbers are composite numbers since they have 2 as a factor. Which of the following are composite numbers since they have 2 as a factor? _____

    (a) 4   (b) 7   (c) 10   (d) 17   (e) 22   (f) 23

    | (a), (c), and (e)

20. The number "3" is a prime number since its only factors are 1 and 3. However, all other multiples of 3 are composite numbers since they have 3 as a factor. Which of the following are composite numbers since they have 3 as a factor? _____

    (a) 5   (b) 9   (c) 12   (d) 17   (e) 21   (f) 29

    | (b), (c), and (e)

21. The number "5" is a prime number since its only factors are 1 and 5. However, all other multiples of 5 are composite numbers since they have 5 as a factor. Which of the following are composite numbers since they have 5 as a factor? _____

    (a) 7   (b) 10   (c) 15   (d) 19   (e) 41   (f) 45

    | (b), (c), and (f)

22. (a) The first three prime numbers are 2, 3, and ___.

    (b) There is one prime number between 5 and 10. It is ___.

    | a) 5   b) 7

23. (a) List the four prime numbers between 10 and 20. _____

    (b) List the two prime numbers between 20 and 30. _____

    | a) 11, 13, 17, and 19   b) 23 and 29

24. Which of the following are prime numbers? _____

    (a) 31   (b) 34   (c) 37   (d) 42   (e) 47   (f) 49

    | (a), (c), and (e)

6-4 THE MEANING OF PRIME FACTORING

In this section, we will discuss what is meant by the "prime" factoring of a number.

25. A factoring is called a "prime" factoring only if each of the factors is a prime number. For example:

    10 = 2 x 5 is a prime factoring of 10 since both 2 and 5 are prime numbers.

    12 = 2 x 6 is not a prime factoring of 12 since 6 is not a prime number.

    Which of the following are prime factorings? _____

    (a) 6 = 2 x 3   (b) 8 = 2 x 4   (c) 15 = 3 x 5   (d) 18 = 3 x 6

    | Only (a) and (c)

26. A factoring of a number into itself and "1" is not a prime factoring since "1" is not a prime number. For example:

    5 = 1 x 5 is not a prime factoring since "1" is not a prime number.

    Which of the following are prime factorings? _____

    (a) 7 = 1 x 7   (b) 9 = 3 x 3   (c) 10 = 2 x 5   (d) 3 = 1 x 3

    Only (b) and (c)

27. A factoring into three or more factors is also called a "prime" factoring only if each of the factors is a prime number. For example:

    12 = 2 x 2 x 3 is a prime factoring of 12 since all three factors are prime numbers.

    16 = 2 x 2 x 4 is not a prime factoring of 16 since 4 is not a prime number.

    Which of the following are prime factorings? _____

    (a) 24 = 2 x 2 x 6   (b) 18 = 2 x 3 x 3   (c) 20 = 2 x 2 x 5   (d) 48 = 2 x 2 x 3 x 4

    Only (b) and (c)

28. Prime numbers can only be factored into themselves and "1". For example:

    3 = 1 x 3   5 = 1 x 5   7 = 1 x 7   13 = 1 x 13

    Since a factoring of a number into itself and "1" is not a prime factoring, there is no prime factoring for prime numbers. Therefore, there is no prime factoring for which of the following numbers?

    (a) 11   (b) 10   (c) 14   (d) 17   (e) 19   _____

    (a), (d), and (e)

29. There is a prime factoring for every composite number. Sometimes the prime factoring contains only two factors. For example:

    6 = 2 x 3   14 = 2 x 7   (a) 10 = 2 x ____   (b) 21 = 3 x ____

    a) 2 x 5   b) 3 x 7

30. Frequently the prime factoring of a composite number contains three or more factors. For example:

    12 = 2 x 2 x 3   40 = 2 x 2 x 2 x 5   (a) 27 = 3 x 3 x ____   (b) 30 = 2 x 3 x ____

    a) 3 x 3 x 3   b) 2 x 3 x 5

31. Though any composite number can be factored in more than one way, there is only one prime factoring for each composite number. Which of the following is the prime factoring for 24? _____

    (a) 24 = 3 x 8   (b) 24 = 4 x 6   (c) 24 = 2 x 2 x 2 x 3   (d) 24 = 2 x 2 x 6

    (c)

32. When writing the prime factoring of a number, we write the factors in order according to their size. The smaller factors are written on the left. For example:

    We write 35 = 5 x 7 instead of 35 = 7 x 5.

    We write 30 = 2 x 3 x 5 instead of 30 = 2 x 5 x 3 or 3 x 2 x 5.

    In which case below are the prime factors of 42 written in the proper order? _____

    (a) 42 = 3 x 7 x 2   (b) 42 = 2 x 3 x 7   (c) 42 = 3 x 2 x 7   (d) 42 = 7 x 3 x 2

    (b)

126

6-5 PRIME FACTORING BY THE SMALLEST-PRIME METHOD

In this section, we will discuss the smallest-prime method for obtaining the prime factoring of a composite number.

33. To obtain the prime factoring of a number with the factors in the proper order, we can use these steps:

    (1) Begin by factoring out the smallest possible prime factor.

    (2) Then continue to factor out the smallest possible prime factor from the factor on the far right until a prime factor is obtained on the far right.

    As an example, we have found the prime factoring of 12 below. Notice these points:

    (1) We began by factoring out "2", the smallest possible prime factor.   12 = 2 x 6

    (2) Then we factored out another "2" from 6, the factor on the far right. Since "3" is a prime factor, the prime factoring was completed.   12 = 2 x 2 x 3

    Following the steps above, complete the prime factoring of 30 at the right.

    30 = 2 x 15

    30 = 2 x ____ x ____

    > 30 = 2 x <u>3</u> x <u>5</u>

34. Using the same method, we have found the prime factoring of 24 below. Notice these points:

    (1) We began by factoring out a "2" from 24.   24 = 2 x 12

    (2) We then factored out another "2" from 12.   24 = 2 x 2 x 6

    (3) We then factored out another "2" from 6. Since "3" is a prime factor, the prime factoring was completed.   24 = 2 x 2 x 2 x 3

    Following the steps above, complete the prime factoring of 36 in the space at the right.

    36 =

    > 36 = 2 x 2 x 3 x 3

35. (a) Find the prime factoring of 8. (b) Find the prime factoring of 20.

    8 =   20 =

    > a) 8 = 2 x 2 x 2
    > b) 20 = 2 x 2 x 5

36. (a) Find the prime factoring of 44. (b) Find the prime factoring of 60.

    44 =   60 =

    > a) 44 = 2 x 2 x 11
    > b) 60 = 2 x 2 x 3 x 5

37. The smallest prime that can be factored out of a number can be larger than "2".

   For example: In the prime factoring of 45, $45 = 3 \times 15$
                 the smallest possible prime is "3". $45 = 3 \times 3 \times 5$

                 In the prime factoring of 125, $125 = 5 \times 25$
                 the smallest possible prime is "5". $125 = 5 \times 5 \times 5$

   (a) Find the prime factoring of 63.   (b) Find the prime factoring of 75.

   63 =   75 =

   a) $63 = 3 \times 3 \times 7$
   b) $75 = 3 \times 5 \times 5$

38. (a) Find the prime factoring of 99.   (b) Find the prime factoring of 100.

   99 =   100 =

   a) $99 = 3 \times 3 \times 11$
   b) $100 = 2 \times 2 \times 5 \times 5$

39. When the prime factoring of a number contains only two factors, we can obtain its prime factoring in one step. For example:

   $4 = 2 \times 2$ $15 = 3 \times 5$ (a) $9 = \underline{\quad} \times \underline{\quad}$ (b) $35 = \underline{\quad} \times \underline{\quad}$

   a) 3×3 b) 5×7

6-6 PRIME FACTORING BY THE FACTOR-TREE METHOD

In this section, we will discuss a second method for finding the prime factoring of a composite number. It is called the "factor-tree" method.

40. There are three steps in the factor-tree method for finding the prime factoring of a composite number. They are:

   (1) Start by factoring the number into any two factors.

   (2) If either factor is not prime, continue factoring it until all factors are prime.

   (3) Then write the prime factors in order according to their size.

   As an example, we have used the factor tree method to find the prime factoring of 30 below. Notice the steps:

   (1) We began by factoring 30 into 5×6.

   (2) Since 6 is not prime, we factored it into 2×3.

   (3) Writing the prime factors in order according to their size, we got $2 \times 3 \times 5$.

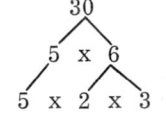

   $30 = 2 \times 3 \times 5$

   Use the factor-tree method to complete the prime factoring of 45 at the right.

   $45 = \underline{\quad} \times \underline{\quad} \times \underline{\quad}$

   $45 = 3 \times 3 \times 5$

41. We used a factor tree to find the prime factoring of 90 at the right.

    Note: Both 9 and 10 had to be factored again since neither is a prime number.

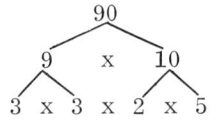

    90 = 2 x 3 x 3 x 5

    Use the factor-tree method to complete the prime factoring of 54 at the right.

    54 = _____ x _____ x _____ x _____

    54 = 2 x 3 x 3 x 3

42. When using the factor-tree method, we can begin by factoring the composite number into any two factors. Therefore, different factor trees can be used to find the prime factoring of some numbers. For example, we have used two different factor trees for the number "60" below.

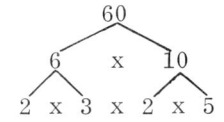

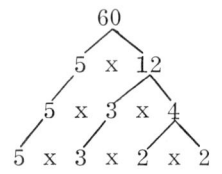

    As you can see, we obtain the same prime factoring for 60 with either tree. That is, 60 = _____ x _____ x _____ x _____ .

    60 = 2 x 2 x 3 x 5

43. Either the "smallest-prime" method or the "factor-tree" method can be used to find the prime factoring of a composite number. Use either method for the problems below.

    (a) Find the prime factoring of 28.   (b) Find the prime factoring of 84.

44. Use either method to complete the problems below.

    (a) Find the prime factoring of 66.   (b) Find the prime factoring of 120.

    a) 28 = 2 x 2 x 7
    b) 84 = 2 x 2 x 3 x 7

45. Remember that the prime factoring of a number can contain only two factors.

    (a) Find the prime factoring of 26.   (b) Find the prime factoring of 77.

    a) 66 = 2 x 3 x 11
    b) 120 = 2 x 2 x 2 x 3 x 5

    a) 26 = 2 x 13
    b) 77 = 7 x 11

SELF-TEST 14 (Frames 1-45)

1. List the next three prime numbers in this sequence: 2, 3, 5, ____, ____, ____

2. Which of the following are <u>prime</u> numbers? (a) 17 (b) 21 (c) 31 (d) 38	3. Which of the following are <u>composite</u> numbers? (a) 23 (b) 24 (c) 39 (d) 43

Find the prime factoring of each of the following:

4. 9 =	5. 50 =	6. 16 =	7. 105 =

ANSWERS: 1. 7, 11, 13   2. (a), (c)   4. 9 = 3 x 3   6. 16 = 2 x 2 x 2 x 2
           3. (b), (c)   5. 50 = 2 x 5 x 5   7. 105 = 3 x 5 x 7

6-7   FINDING LOWEST COMMON DENOMINATORS BY THE "PRIME-FACTORING" METHOD

We showed earlier that it is sometimes difficult to find the lowest common denominator for an addition or subtraction by the "direct" method. In those cases, we can use another method called the "prime-factoring" method. We will discuss the "prime-factoring" method in this section.

46. We can find the lowest common denominator for $\frac{1}{6} + \frac{5}{21}$ by the "<u>direct</u>" method. That is, checking multiples of 21 until we find the lowest one that is also a multiple of 6, we get 42 as the lowest common denominator.

There is another method called the "<u>prime-factoring</u>" method that can be used to find the lowest common denominator for the same addition. The two steps in the "prime-factoring" method are shown and described below.

<u>Step 1</u>   <u>Step 2</u>

6 = 2 x 3   2   3
21 = 3 x 7   3   7
              ─────────
              2 x 3 x 7 = 42

Step 1: We wrote the prime factoring of each denominator.

Step 2: (a) We lined up the prime factors of the two denominators so that the common factor "3" was in the same column.

(b) We wrote a multiplication of prime factors "2 x 3 x 7" in which the common factor "3" was used only once.

(c) We performed the multiplication 2 x 3 x 7 and got 42.

Is 42 the same lowest common denominator that we obtained by the "direct" method? _____

Yes

130

47. Using the "direct" method, we get 30 as the lowest common denominator for $\frac{5}{6} + \frac{1}{10}$. The steps in the "prime-factoring" method of finding the lowest common denominator for the same addition are shown below.

Step 1   Step 2

6 = 2 x 3   2 3
10 = 2 x 5   <u>2   5</u>
              2 x 3 x 5 = 30

Note: In Step 1, we found the prime factoring of each denominator.

In Step 2, we lined up the common factor "2" in the same column and used it only once in the multiplication "2 x 3 x 5".

Using the "prime-factoring" method, we got 30 as the lowest common denominator. Is this the same number obtained by the "direct" method? _____

48. Using the direct method, we get 60 as the lowest common denominator for $\frac{5}{12} - \frac{4}{15}$. The steps in the "prime-factoring" method are shown below.

Step 1   Step 2

12 = 2 x 2 x 3   2 2 3
15 = 3 x 5   <u>   3 5</u>
                  2 x 2 x 3 x 5 = 60

Notice these two points about the multiplication "2 x 2 x 3 x 5" in Step 2.

(1) The factor "2" is used twice since it appears twice in the prime factoring of 12.

(2) The common factor "3" is used only once.

Using the prime-factoring method, we got 60 as the lowest common denominator. Is this the same number obtained by the direct method? _____

49. Using the direct method, we get 84 as the lowest common denominator for $\frac{5}{12} + \frac{11}{42}$. The steps in the prime-factoring method are shown below.

Step 1   Step 2

12 = 2 x 2 x 3   2 2 3
42 = 2 x 3 x 7   <u>   2 3 7</u>
                  2 x 2 x 3 x 7 = 84

Notice these points about the multiplication "2 x 2 x 3 x 7":

(1) "2" is used as a factor twice since it appears twice in the prime factoring of 12. The "2" from the factoring of 42 is not used since it is common.

(2) The common factor "3" is used only once.

Using the prime-factoring method, we got 84 as the lowest common denominator. Is this the same number obtained by the direct method? _____

Answers:
47. Yes
48. Yes
49. Yes

50. The prime-factoring method for $\frac{3}{8} - \frac{5}{36}$ is shown at the right.

    Step 1
    8 = 2 x 2 x 2
    36 = 2 x 2 x 3 x 3

    Step 2
    2 2 2
    _ 2 2 3 3_
    2 x 2 x 2 x 3 x 3

    Notice these points about "2 x 2 x 2 x 3 x 3":

    (1) "2" is used as a factor three times since it appears three times in the prime factoring of 8. The two 2's from the prime factoring of 36 are not used since they are common.

    (2) "3" is used as a factor twice since it appears twice in the prime factoring of 36.

    Perform the multiplication at the right in order to find the lowest common denominator.   2 x 2 x 2 x 3 x 3 = _____

---

51. The prime-factoring method for $\frac{5}{24} + \frac{7}{32}$ is shown at the right.

    Step 1
    24 = 2 x 2 x 2 x 3
    32 = 2 x 2 x 2 x 2 x 2

    Step 2
    2 2 2   3
    _2 2 2 2 2_
    2 x 2 x 2 x 2 x 2 x 3

    Notice these points about "2 x 2 x 2 x 2 x 2 x 3":

    (1) "2" is used as a factor five times since it appears five times in the prime factoring of 32. The three 2's from the factoring of 24 are not used since they are common.

    (2) "3" is used as a factor once since it appears once in the prime factoring of 24.

    Perform the multiplication at the right to find the lowest common denominator.   2 x 2 x 2 x 2 x 2 x 3 = _____

72

---

52. The prime-factoring method can be used when the denominators do not contain a common factor. As an example, we have used that method to find the lowest common denominator for the addition at the left below.

    $\frac{3}{4} + \frac{5}{21}$

    Step 1
    4 = 2 x 2
    21 = 3 x 7

    Step 2
    2 2
    _   3 7_
    2 x 2 x 3 x 7

    Therefore, the lowest common denominator for the addition above is _____.

96

---

53. Even though prime numbers have no prime factoring, we can use the prime-factoring method when one denominator is a prime number. As an example, we have used that method for the subtraction at the left below. Notice that the denominator "5" is one of the factors in "2 x 2 x 3 x 5".

    $\frac{4}{5} - \frac{5}{12}$

    Step 1
    5 = 5
    12 = 2 x 2 x 3

    Step 2
              5
    _2 2 3_
    2 x 2 x 3 x 5

    Therefore, the lowest common denominator for the subtraction above is _____.

84

60

131

132

6-8 A SHORTER FORM OF THE PRIME-FACTORING METHOD

There is a shorter form of the prime-factoring method in which the multiplication used to find the lowest common denominator is written directly from the two prime factorings. We will discuss the shorter form in this section.

54. In the prime-factoring method, a multiplication of prime factors is used to find the lowest common denominator. To obtain this multiplication, we use the following general rules:

   (1) Each different prime factor must appear.

   (2) Each prime factor must appear as many times as it appears in the prime factoring where it occurs the greatest number of times.

   (3) Common factors are not repeated.

   By following these rules, we can write the multiplication directly from the prime factorings. An example is given below.

   To find the lowest common denominator for $\frac{4}{15} + \frac{5}{21}$, $15 = 3 \times 5$
   $21 = 3 \times 7$
   we have written the prime factorings of 15 and 21 at the right. Let's write the required multiplication in steps by examining the factors.

   (1) "3" must appear once since it appears once in both factorings.   3

   (2) "5" must appear once since it appears once in the factoring of 15. 3×5

   (3) "7" must appear once since it appears once in the factoring of 21. $3 \times 5 \times 7$

   Perform the final multiplication to find the lowest common denominator for the addition above.

55. To find the lowest common denominator for $\frac{3}{8} - \frac{5}{14}$, $8 = 2 \times 2 \times 2$
   $14 = 2 \times 7$
   we have written the prime factorings of 8 and 14 at the right. Let's write the required multiplication in steps by examining the factors.

   (1) "2" must appear three times since it appears three times in the factoring of 8. $2 \times 2 \times 2$

   (2) "7" must appear once since it appears once in the factoring of 14. $2 \times 2 \times 2 \times 7$

   Perform the final multiplication to find the lowest common denominator.

105

56

133

56. To find the lowest common denominator for $\frac{11}{12} + \frac{5}{18}$, we have written the prime factorings of 12 and 18 at the right. Let's write the required multiplication in steps by examining the factors.

    12 = 2 x 2 x 3
    18 = 2 x 3 x 3

    (1) "2" must be used twice since it appears twice in the factoring of 12.

    2 x 2

    (2) "3" must be used twice since it appears twice in the factoring of 18.

    2 x 2 x 3 x 3

    Perform the final multiplication to find the lowest common denominator.

---

57. To find the lowest common denominator for $\frac{13}{20} - \frac{7}{24}$, we have written the prime factorings of 20 and 24 at the right. Let's write the required multiplication in steps by examining the factors.

    36

    20 = 2 x 2 x 5
    24 = 2 x 2 x 2 x 3

    (1) "2" must appear three times since it appears three times in the factoring of 24.

    2 x 2 x 2

    (2) "3" must appear once since it appears once in the factoring of 24.

    2 x 2 x 2 x 3

    (3) "5" must appear once since it appears once in the factoring of 20.

    2 x 2 x 2 x 3 x 5

    Perform the final multiplication to find the lowest common denominator.

---

58. For each addition below, we have written the prime factorings of the two denominators. Write the multiplication of factors needed in each case to find the lowest common denominator.

    120

    (a) $\frac{9}{10} + \frac{3}{14}$ 10 = 2 x 5
                14 = 2 x 7

    (b) $\frac{1}{6} + \frac{7}{16}$ 6 = 2 x 3
                16 = 2 x 2 x 2 x 2

---

59. For each subtraction below, we have written the prime factorings of both denominators. Write the multiplication of factors needed in each case to find the lowest common denominator.

    a) 2 x 5 x 7
    b) 2 x 2 x 2 x 2 x 3

    (a) $\frac{9}{14} - \frac{5}{12}$ 12 = 2 x 2 x 3
                14 = 2 x 7

    (b) $\frac{7}{9} - \frac{1}{21}$ 9 = 3 x 3
                21 = 3 x 7

---

60. Following the steps below, use the prime-factoring method to find the lowest common denominator for $\frac{3}{8} + \frac{5}{18}$.

    a) 2 x 2 x 3 x 7
    b) 3 x 3 x 7

    (a) Write the prime factorings of 8 and 18.

    8 = _____
    18 = _____

    (b) Write the multiplication of factors needed to find the lowest common denominator.

    _____

    (c) Perform the multiplication to get the lowest common denominator.

    _____

61. Following the steps below, use the prime-factoring method to find the lowest common denominator for $\frac{11}{12} - \frac{7}{10}$.

    (a) Write the prime factorings of 10 and 12.

    10 = _____
    12 = _____

    (b) Write the multiplication of factors needed to find the lowest common denominator.

    _____

    (c) Perform the multiplication to get the lowest common denominator.

    _____

a) 8 = 2 x 2 x 2
    18 = 2 x 3 x 3
b) 2 x 2 x 2 x 3 x 3
c) 72

62. Using the same steps, find the lowest common denominator for each addition below.

    (a) $\frac{3}{10} + \frac{1}{16}$ _____

    (b) $\frac{11}{30} + \frac{5}{18}$ _____

a) 10 = 2 x 5
    12 = 2 x 2 x 3
b) 2 x 2 x 3 x 5
c) 60

63. Since the "direct" method is faster than the "prime-factoring" method, <u>only</u> use the <u>prime-factoring</u> method <u>in cases when the direct method is difficult to use.</u>

Use either method to find the lowest common denominator for each subtraction below.

    (a) $\frac{3}{40} - \frac{1}{80}$ _____

    (b) $\frac{11}{12} - \frac{7}{20}$ _____

a) 80, from 2 x 2 x 2 x 2 x 5
b) 90, from 2 x 3 x 3 x 5

64. Use either method to find the lowest common denominator for each addition below.

    (a) $\frac{5}{16} + \frac{3}{14}$ _____

    (b) $\frac{1}{9} + \frac{7}{15}$ _____

a) 80   b) 60

a) 112   b) 45

6-9 THE PRIME-FACTORING METHOD AND ADDITIONS OF THREE FRACTIONS

The prime-factoring method for identifying lowest common denominators is especially useful for many additions of three fractions. We will discuss its use with additions of three fractions in this section.

65. To find the lowest common denominator for $\frac{2}{9} + \frac{5}{8} + \frac{7}{12}$, we have written the prime factorings of 9, 8, and 12 at the right. Let's find the multiplication of factors needed in order to find the lowest common denominator. Two steps are needed:

$9 = 3 \times 3$
$8 = 2 \times 2 \times 2$
$12 = 2 \times 2 \times 3$

(1) "2" is used three times since it appears three times in the prime factoring of 8.

$2 \times 2 \times 2$

(2) "3" is used twice since it appears twice in the prime factoring of 9.

$2 \times 2 \times 2 \times 3 \times 3$

Perform the final multiplication to find the lowest common denominator.

66. For each addition below, we have written the prime factorings of each denominator. Write the multiplication of factors needed to find the lowest common denominator in each case.

72

(a) $\frac{1}{6} + \frac{1}{10} + \frac{1}{16}$ $6 = 2 \times 3$
$10 = 2 \times 5$
$16 = 2 \times 2 \times 2 \times 2$ _____

(b) $\frac{1}{10} + \frac{1}{14} + \frac{1}{18}$ $10 = 2 \times 5$
$14 = 2 \times 7$
$18 = 2 \times 3 \times 3$ _____

67. Let's use the prime-factoring method to find the lowest common denominator for $\frac{3}{14} + \frac{10}{21} + \frac{7}{30}$.

a) $2 \times 2 \times 2 \times 2 \times 3 \times 5$
b) $2 \times 3 \times 3 \times 5 \times 7$

(a) Write the prime factorings of 14, 21, and 30.
$14 = $ _____
$21 = $ _____
$30 = $ _____

(b) Write the multiplication needed to find the lowest common denominator.   _____

(c) Perform the multiplication to find the lowest common denominator.   _____

68. Use the prime-factoring method to find the lowest common denominator in each case below.

(a) $\frac{3}{4} + \frac{5}{6} + \frac{3}{14}$ ____   (b) $\frac{1}{12} + \frac{1}{18} + \frac{1}{20}$ ____

a) $14 = 2 \times 7$
$21 = 3 \times 7$
$30 = 2 \times 3 \times 5$
b) $2 \times 3 \times 5 \times 7$
c) 210

135

69. We used the prime-factoring method to find the lowest common denominator for $\frac{2}{7} + \frac{9}{10} + \frac{8}{15}$ at the right.

$7 = 7$
$10 = 2 \times 5$
$15 = 3 \times 5$

$2 \times 3 \times 5 \times 7 = 210$

a) 84, from $2 \times 2 \times 3 \times 7$
b) 180, from $2 \times 2 \times 3 \times 3 \times 5$

Note: Though 7 is a prime number and has no prime factoring, we had to include 7 as a factor in the multiplication.

Use the prime-factoring method to find the lowest common denominator for each addition below.

(a) $\frac{1}{3} + \frac{7}{10} + \frac{8}{25}$ _____

(b) $\frac{1}{5} + \frac{4}{15} + \frac{7}{18}$ _____

---

70. Even with additions of three fractions, <u>only use the prime-factoring method in those cases in which the direct method is difficult to use.</u>

a) 150, from $2 \times 3 \times 5 \times 5$
b) 90, from $2 \times 3 \times 3 \times 5$

Use either method to find the lowest common denominator for each addition below.

(a) $\frac{7}{20} + \frac{53}{80} + \frac{39}{40}$ _____

(b) $\frac{2}{3} + \frac{3}{8} + \frac{9}{10}$ _____

a) 80   b) 120

SELF-TEST 15 (Frames 46-70)

In each problem below: (a) Find the lowest common denominator.
(b) Then find the sum or difference.

1. $\frac{1}{22} + \frac{1}{8} =$

    (a) _____ (b) _____

2. $\frac{5}{21} - \frac{3}{28} =$

    (a) _____ (b) _____

3. $\frac{1}{5} + \frac{1}{18} + \frac{1}{6} =$

    (a) _____ (b) _____

4. $\frac{3}{8} + \frac{5}{12} + \frac{1}{30} =$

    (a) _____ (b) _____

ANSWERS:

1. (a) 88   (b) $\frac{15}{88}$ 2. (a) 84   (b) $\frac{11}{84}$ 3. (a) 90   (b) $\frac{19}{45}$ 4. (a) 120   (b) $\frac{33}{40}$